Raising an organisation's Collective IQ

Onno van Ewyk is a management consultant who has written extensively on management practices. His expertise lies in knowledge management, quality management, and the development of effective management systems. In short, the key aspects of organisational activity that build sustainable profitability and growth.

Preface to the Second Edition

This edition updates the first edition in several ways. Firstly, and most importantly, the term "Business Information Topic" has been changed to "Guidance Topic".

A Guidance Topic is a document that combines policy, procedure, and instructional information to meet a single knowledge need. It overcomes the problems and limitations of separately labelling documents as 'Policy', 'Procedure', 'Instruction', or other such term. This is a key concept in a critique of current organisational approaches to documentation, which is explained in detail in the first chapter titled *An atomic unit of knowledge*.

While the term Business Information Topic has appeal in the sense that 'business information' conveys a close connection to an organisation's operations (another key concept explored in Chapter 2 *Dynamically picturing the organisation*) and it has the snappy acronym BIT, it does not sufficiently convey the essential notions of being useful, helpful, and supportive. Hence Guidance Topic as a replacement.

Some may argue that 'guidance' is too 'soft' a term and that some documents should be called 'policy', or 'standard operating procedure', or such like, to impress on the reader that they are compelled to comply with its contents. But it can also be argued that in practice words cannot control behaviour, they can only influence it. And they can only influence it effectively by being relevant, accessible, and easily understood, which is precisely what a Guidance Topic aims to be.

Other revisions include:

- an additional appendix that takes a deep dive into how to write a good Guidance Topic
- an update to the SharePoint illustrations - these have been changed to the 'modern' interface (from the older 'classic' interface)
- slight changes to other illustrations to make them easier to read

And, of course, there have been corrections to some typographical errors and minor omissions.

Onno van Ewyk

Raising an Organisation's Collective IQ

All inquiries should be made to the author at ovanewyk@veco.net.au or PO Box 293, Rozelle NSW Australia 2039.

ISBN: 978-0-6456308-0-0

NATIONAL LIBRARY OF AUSTRALIA

A catalogue record for this work is available from the National Library of Australia

Published by FiveZero Press Sydney Australia

5z Press

Cover art & design by Charlotte Creature

Acknowledgement

This book draws extensively on a conceptual framework developed by the late Dr Douglas Engelbart, the digital pioneer who initiated the networked personal computing revolution. In 1968 he demonstrated for the first time the potential for a computer to be both a personal tool and a virtual environment in which people can collaborate on a previously unimagined scale[1]. His motive was not commercial gain or personal fame but the betterment of humanity.

I was privileged to meet Dr Engelbart at a conference at which he was the keynote speaker and at which I was presenting a paper on knowledge management. I was inspired by his ideas. He was kind enough to attend my presentation and remark on its compatibility with his ideas.

Despite the briefness of our encounter, I remember him as a thoughtful, modest, and patient man. I hope this book will, in some small way, help to further advance his vision of "augmenting the human intellect".

---

[1] "Engelbart gave a demonstration of the "oN-Line System" at the Fall Joint Computer Conference in San Francisco on Dec. 9, 1968. The oN-Line System was the first hypertext system, preceding the web by more than 20 years." - Wired Magazine September 2018

Contents

Introduction

The aim of this book is to challenge conventional concepts on how to document, disseminate, and develop knowledge within organisations. These concepts, combined with a general lack of recognition of the value of well-constructed and integrated knowledge sharing systems, prevents organisations from taking full advantage of digital environments to improve their performance.

The alternative concepts, and the techniques associated with them, enable employees to work more effectively as individuals, as team members, as team leaders, and as managers for the overall betterment of the organisation. And they can be implemented without significant expenditure on additional resources in terms of people, systems, software, or equipment. The required resources are already in place – they just need to be utilised with a different mindset.

The core of the problem with current approaches to documented knowledge in organisations is that they are typically ad-hoc and inconsistent. This chokes off their potential benefits and even has negative impacts. This is particularly the case for documented knowledge on how and why things are done - so called "policies and procedures". The phrase itself has an ad-hoc 'bucket of stuff' connotation that limits the way we think about it.

Many managers see this documentation as a chore or an afterthought that is required only to satisfy auditors and regulators. They completely miss its potential to be a powerful tool for achieving consistent performance, stimulating innovation, and facilitating change when combined with today's technology.

The first step towards realising this potential is to replace Policies and Procedures with a different concept for encapsulating and communicating knowledge; one that reflects contemporary views on how best to communicate in writing and that takes full advantage of the electronically connected environment in which we all now work. This is the subject of the chapter titled *An atomic unit of knowledge*, which draws on research carried out in the fields of knowledge management and technical communication to create a radically better approach to documentation.

The immediate benefit of this approach is easier and better access to knowledge for individuals. It also helps to improve the quality of knowledge by making it easier to create and maintain.

Quality in this case means useful, accurate, comprehensive, up-to-date, and authoritative. There is some complexity to ensuring that these criteria are met but the key is to manage it within a framework of process control and improvement. This is the subject of the chapter titled *Dynamically picturing the organisation*, which connects the 'value chain' concept and quality management ideas to managing documentation.

Many organisations apply quality management and process improvement principles to very good effect, but most often this is in selective or sporadic ways. Raising Collective IQ requires a whole-of-organisation approach that is integrated with knowledge management.

Managers are key to this. They need to see policies and procedures as process knowledge and therefore as a means of leveraging consistent and improved performance for their teams. And as a means of leveraging their team members' feedback, ideas, and innovations.

Managing process knowledge in a dynamic way can only be done using computer networks and the right software to manage documents electronically. This is the subject of the chapter titled *Creating the right virtual environment*. Computer networks and electronic document management are the enabling technologies for raising Collective IQ.

The problem is that much of what is done digitally is carried over from the non-digital past. This means that the opportunity to manage knowledge in a dynamic way, which includes real-time editing, effective classification schemes, and full exploitation of hypertext capabilities, goes begging - and performance suffers as a result.

Taking the actions recommended in these first three chapters will go a long way towards raising the organisation's Collective IQ. To get to the next level requires some reflection on why digital capabilities exist in the form that they do today. This means acknowledging the work of the digital pioneer who coined the term 'Collective IQ' – Doug Engelbart. This is the subject of the chapter titled *Doug Engelbart and Collective IQ*.

Engelbart identified five key concepts behind raising Collective IQ:

- Networked Improvement Community (NIC)
- Dynamic Knowledge Repository (DKR)
- Concurrent Development Integration and Application of Knowledge (CoDIAK)
- Open Hyperdocument System (OpHys)
- Bootstrap

In his vision, the application of these concepts enhances society's ability to solve complex, large-scale problems. The aim of this book is to apply these concepts to organisations to enhance their capabilities to deliver quality products and services to customers, to support innovation and adaptation, and to enrich their employees' work experience.

In the chapters that follow, the focus is on Dynamic Knowledge Repositories (DKRs) and Concurrent Development Integration and Application of Knowledge (CoDIAK) and the complexities involved in developing these to transform cultures from an implicit 'command-and-control' mentality to an explicit one of 'knowledge sharing'.

Collective IQ would not represent a true benefit to society were it not value driven. In this case the core value is quality – quality in the sense of a commitment by the organisation to creating, maintaining, and improving customer value. As a concept it is a natural fit with continuous process improvement and effective knowledge management.

This means a genuine cultural shift needs to take place, particularly in the C suite. A cultural shift that moves an organisation's primary focus away from short term profits and trusts that healthy and sustainable profits and shareholder value flow naturally from consistently delivered customer value.

Moving towards a culture biased towards knowledge sharing and quality, and away from one biased towards command-and-control and short-term profit maximisation also requires commitment to sincerity and integrity. This is because the currency of Collective IQ is documented knowledge, which if ignored, bypassed, or neglected becomes devalued, often with devastating effects. This is the subject of the penultimate chapter titled *Combatting organisational stupidity*.

Having started at a micro level and progressively expanded the view to cover the whole organisation and its place in society, the chapter titled

Intelligent free enterprise – enhancing the human intellect discusses Engelbart's final two components of Collective IQ, Open Hyperdocument System (OpHys) and Bootstrap. These are concepts that go beyond the boundaries of individual organisations and promise problem-solving capabilities at a much larger scale.

1. An atomic unit of knowledge

The way an organisation functions determines how successful it will be. The way objectives and values are determined, and how they are commonly understood, directly affect the way employees behave and the day-to-day decisions that they make.

The organisation's methods and processes, including those that are automated, determine the quality of products and services, the efficiency with which they are made and delivered, and ultimately the organisation's profitability and sustainability.

All these things – objectives, values, methods, and processes – are held as knowledge, both in the minds of those working in the organisation, and in documents that people can consult when in doubt, when they need training, or when they need to instruct others.

The potential for these documents to benefit the organisation is currently only fractionally realised because the way they are created and managed reflect outdated views on how organisations work and fail to take advantage of now ubiquitous digital work environments.

The core documents concerned are commonly labelled 'policy', 'procedure', 'work instruction', 'manual' and the like, and they are generally managed in a fragmented and ad-hoc way to support specific areas of activity. Production functions maintain 'Standard Operating Procedures', Human Resources functions publish 'Employee Manuals', and so on. These are invariably created on the assumption of 'one way' communication – from above to below – with no real opportunity for interaction or dialogue around their contents. This severely limits their potential to dynamically improve the organisation's capabilities.

The problem is that the concepts of 'policy', 'procedure' and 'work instruction' as discrete documents are relics of a world of paper-publishing. Documents were typically bundled into ring-binders, updated sporadically and often widely ignored because they were difficult to access (because of limited numbers of copies) and often out of date because it was cumbersome and logistically difficult to update them.

The language used was also often problematic. There was a general belief that what was written down and published should sound important and 'official', but this often resulted in a tone that was bureaucratic, stiff, and remote. Readers struggled to connect with it and to clearly understand what it meant for them and their work.

These shortcomings led to a high reliance on word-of-mouth advice, but word-of-mouth requires direct interaction, which is expensive in terms of staff time, and the information conveyed can be parochial or idiosyncratic or vary depending on who is asked.

In the absence of personal advice, people resort to trial and error (aka, reinventing wheels). The problem of course is that 'trial' is invariably time-consuming and 'error' is wasteful, expensive, and potentially catastrophic.

The networked digital environment that now prevails has the potential to overcome these shortcomings but organisations often fail to take advantage. Documents that were once distributed on paper are simply converted into PDFs, with the same use of bureaucratic language, the same ineffective writing style, and the same inflexibility. Here are the tell-tale signs of this anachronistic approach:

- PDFs with little or no use of hypertext links (usually an indication that content is developed in isolation and published sporadically)
- scanned hand-written signatures (despite almost every employee now having an online identity that is much harder to fake than a scanned signature)
- 'Document Control' tables that are manually updated in the document (despite the ready availability of content control systems that include version histories)
- file names that contain version numbers (again, despite the ready availability of content control systems that automatically assign version numbers, which can be made to appear in the document and update automatically)
- documents that begin with a separate title page that must be scrolled out of the way to get to the content
- documents with standardised headings, which, under the pressure of filling an otherwise blank space, result in content that is repetitive, generic, and, in many cases, redundant
- long documents that are subject-based rather than short ones focused on users' needs and circumstances

- documents in Intranets grouped under the headings of 'Policies' and 'Procedures', instead of the categories of organisational activity to which they relate

The last point about classifying and characterising documents is particularly important because it has an influence on how documents are written. The reason for this is that the words 'policy' and 'procedure' (and others such as 'work instruction' or 'standard operating procedure'), implicitly assume that businesses are managed on a simplistic 'command-and-control' basis.

Deconstructing the document pyramid

Organisations do not operate on a strict command-and-control basis, and yet the concept persists in the way this key internal information is communicated through the widely promoted concept of a 'pyramid' of information. According to this conventional wisdom, organisational information divides itself logically into three distinct levels - policies at the top, procedures next, and work instructions at the bottom. These exist in a strict hierarchy, which is typically represented as a pyramid (see Figure 1).

Figure 1 Deconstruct the document pyramid

Policies are at the peak, establishing overall aims and values, and held as the preserve of the Board and top management. Procedures then sit under policies in the next tier of the pyramid and explain how policies are put into practice via detailed processes. These are the preserve of middle management, who are charged with putting policies into practical effect.

At the bottom of the pyramid are work instructions. These are handed out to workers or front-line staff to provide step by step instructions for completing individual tasks. In this model, senior managers and the Board are viewed as all-seeing and all-knowing sources of wisdom, middle managers are dutiful lieutenants who convert the wisdom into action, and front-line workers are compliant foot soldiers who follow orders explicitly.

This structure has a superficial appeal. It has a certain neat and tidy look that mirrors a perceived disciplined command-and-control environment that flows majestically from top to bottom, from wood-panelled board room to busy shop front or assembly line. But it is, in fact, nonsense. Organisations simply do not work in such a neat and tidy hierarchical way. Organisational activity is messy, demanding, and constantly permeated with problems, changes, doubts, lessons, contingencies, and adaptations. This is why there is now so much emphasis placed on teamwork and on organisational relationships that are flexible and adaptable, which means that the way documented knowledge is handled needs to follow suit to be effective.

Facing reality

For documented information to be useful to someone trying to get their work done, or make a sensible decision, it needs to address that person's circumstances and contain all the information components relevant to acting in those circumstances. These components are highly likely to include a *combination* of policy, procedural, and instructional information.

For example, information on how to prepare advertising copy may cover:

- who can write it in terms of pre-requisite skills and designated responsibility (policy)
- the methodology to follow (instruction)

- compliance with legal and industry code requirements (policy)
- cross-references /links to compliance information (optional procedure contingent on personal knowledge)
- formats or forms to use to capture all required information for review and approval (procedure)
- who reviews and approves for release (policy) and by what means (work instructions).

What this example highlights, is that the orientation of the traditional approach to this sort of corporate information is wrong. It is prepared from a perspective that pays little explicit attention to who is meant to act on the information and how they are meant to act.

It also often lacks interpretation in terms of a business's specific needs and priorities, and therefore it becomes disconnected and ineffective. Policy documents become generalised motherhood statements, procedures are designed without connections to meaningful policy commitments, and instructions end up being conveyed rigidly without the context or meaning that allows people to adapt to varying needs and situations.

Creating atomic units of knowledge

The essential problem is that policies, procedures, and instructions describe different types of information, not discrete documents. What is needed is a different concept and name for a 'unit of information' that conveys corporate information of this nature, and a different model for structuring these units of information into a meaningful framework (this is the subject of the next chapter).

If the basic orientation of these documents is changed from 'command-and-control' to 'knowledge support' then the way the information is packaged takes a quite different form. Employees working in teams who are empowered to act in the best interests of their customers are best supported by information that is directly relevant to them and easy to understand and apply. This means that the information works best if it is:

- user-focused and easy to access
- confined to a specific issue, decision-point, or task (combining, where applicable, elements of policy, procedure, instruction, and explanation)

- written in a way that is concise, direct, and straightforward (so that it is quickly assimilated to support action or decision-making)
- part of a connected network of inter-related units of information that can be explored in one direction to gain broader understanding, or in the other to focus in on greater detail, depending on the links followed (that is, it should be 'atomic' in the sense that is a building block for more complex information structures)

The new 'atomic unit of knowledge' with these characteristics needs a name or label to support its use as a fresh form of information currency. For this, organisations can leverage a concept developed and refined by the technical communication profession. As part of their transition to increasing digital distribution and management of technical information, they coined the term 'topic', which is defined as:

"A title and content that is short enough to be specific to a single subject or answer a single question, but long enough to make sense on its own and be authored as a unit".

To highlight both its user-focus and its aim to provide immediately useful knowledge to support organisational activity, we can add the word 'guidance' to coin the term 'Guidance Topic' to give a name to this 'atomic unit of knowledge' for organisations.

Replacing policy, procedure, and work instruction documents with Guidance Topics not only breaks down outdated cultural influences but knowledge resources can be built that are much better suited to digital communication and interaction.

Raising Collective IQ using Guidance Topics

Individual managers can begin to raise the Collective IQ of their teams immediately without having to be part of an organisation-wide initiative simply by abandoning conventional approaches to writing policies and procedures and instead use the topic-based approach described above.

This means:

- changing from a mindset of command-and-control to one of knowledge-support or knowledge-sharing

- writing in topic-style to address specific tasks or circumstances and how best to handle them (by combining all the relevant elements of policy, procedure, and instruction)
- employing a writing style that is easy to understand and engage with
- developing topic content collaboratively with team members
- enabling and actively encouraging team members to provide feedback, criticism, and improvement ideas
- using documented information as a basis for common understanding, coordinated action, and collective improvement[2]

Making these changes immediately improves the capacity for written information to bind and focus a team. The next step in raising Collective IQ is to place the Guidance Topics in a commonly understood and meaningful framework, which is the subject of the next chapter.

A note on the importance of plain writing

The point form list above notes the importance of employing a writing style that is easy to understand and engage with. It is worthwhile here expanding on this to highlight the close connection between Collective IQ and the widely employed concept of plain writing (also referred to as 'plain language', 'plain English', or simply 'clear language' or 'clear writing').

Plain writing emerged as a concept in the latter part of the 20th century as a reaction to the style of writing typically employed in Government and business documents – such as regulations, legislation, contracts, insurance and finance forms, social service forms, instructions for interacting with Government agencies, and so on. There was a general consensus that these documents were unnecessarily difficult to understand because of the way they were written.

---

[2] The idea of knowledge packaged as an 'atomic unit' was first suggested by Michael H. Zack (Zack 1999), who stated "… knowledge-as-object becomes knowledge-as-process. The basic structural element is the knowledge unit, a formally defined, atomic packet of knowledge content that can be labelled, indexed, stored, retrieved, and manipulated."

In 1983 the Australian Government implemented the *Plain English and Simpler Forms Program* to improve the quality of writing. This initiative continues now via the *Australian Government Style Guide* sections titled *Clear language and writing style* and *Plain language and word choice*. Also, the *Office of Parliamentary Counsel* publishes a *Plain English Manual* for use in drafting easier-to-understand legislation, and the NSW State Government publishes the *NSW Premier and Cabinet Plain English Guide*.

The United States Government adopted plain writing principles even more forcefully. In 2010 it passed the Plain Writing Act, which mandates the use of plain language in all written interactions between US Government Agencies and the public.

Unfortunately, plain writing principles have had little impact on the way organisations write policies, procedures, and instructions. In fact, the 'document pyramid' orthodoxy indirectly encourages the opposite of plain writing, which is not fancy writing but simply poor writing. It encourages writing that is:

- subject- oriented rather than reader-oriented, which makes it harder to understand and apply.
- written in the passive voice, which avoids identifying responsibility and action when it should be highlighting it
- loaded with bureaucratic expressions and jargon that obscures meaning and alienates the reader

This comes about because the document pyramid separates policies, procedures, and instructions into different documents which immediately hinders the writer's capacity to package knowledge in a way that best meets the reader's (and the organisation's) needs. When combined with the command-and-control viewpoint also implicit in the pyramid, the result is often a sort of high-handed bureaucratese that is unpleasant to read and difficult to apply.

Here are some examples.

- This extract is from a document titled *Standard Operating Procedure HSEQ Element 7 Compliance Procedure Supplier and Contractor Management* published internally by a large mining company.

> *"Service providers play a key role in the [name of organisation] organisation. HSEQ engagement and expectation setting with service providers are critical activities necessary to align the commitment of service providers and their employees and subcontractors to [name of organisation]'s HSEQ vision and values.*
>
> *This procedure specifies the mandatory requirements that must be met to ensure that effective controls are in place for the management of contractors and the HSEQ risks associated with the use of contractors are effectively managed."*

Translated into plain writing, this is simply an instruction to managers when they engage suppliers or contractors that can be restated as:

> *"When you engage a supplier or contractor, make sure that they can meet our requirements for quality, and for protecting health & safety and the environment. To achieve this, take the follow actions."*

Further, the above does not even need a separate document. It can simply be included in a broader topic titled *Engaging suppliers or contractors*. In contrast with the example quoted, this title simply and clearly identifies what the document is about. There is no need to call it a 'Procedure' or even more pompously a 'Standard Operating Procedure'. Ironically it contains policy information (a mandated requirement) as well as procedural information (what action to take).

- This extract is from a document titled *Compliance Risk, Incident & Breach Reporting Policy* published internally by an Australian Bank. It is an example of the rather confused bureaucratic fluff that results when the author is not clear who they are addressing and for what purpose.

> *"The objective of this policy is to articulate the Bank's position on compliance and incident reporting. The Board notes that compliance starts at the top and will be most effective in a culture that emphasises standards of honesty and integrity. The Board of Directors and Executive Management team lead by example and regard compliance*

as a core risk management activity within the Bank. It is acknowledged that it concerns everyone at the Bank and is viewed as an integral part of the Bank's business activities. All operational Managers are responsible for ensuring compliance within their areas of responsibility including supervision and monitoring of all staff in their teams."

None of these words mean anything that will make a difference to the way this bank operates. Risk, incident, and breach reporting are not optional for a Bank. This document should be a Guidance Topic published by the Board that precisely instructs senior management in how they want risk, incident, and breach information collected and reported to them so that they can act on it and ensure compliance.

The last sentence above illustrates how nonsensical this 'policy' document is. Translated into plain writing, it says "Managers are responsible for managing."

- This extract is from a document published by the multi-national health products company RB on their web site. The document is titled *Speak Up Policy Concerning Potential Breaches of Code of Conduct.*

  "The Company would always encourage employees to raise concerns internally in the first instance with their direct line manager or, in the event that this is inappropriate, a member of the Legal, Compliance or HR teams, or another senior person in the business. However, it is not a prerequisite that this occur (sic) before an employee may use one of the other avenues described below for reporting a concern."

  Notice how vague and convoluted the quoted text is. It opens with the passive and conditional phrase "The company would always encourage …" – would if what? It does not say. Then it follows with a further condition "… in the event that this is inappropriate …" but it does not say what circumstances might make it "inappropriate".

  The document is about reporting serious misconduct (not simply 'concerns'), which can be done either internally (to a manager) or externally (such as to an anti-corruption body, the police, an ombudsman, the human rights commission, etc.). One way to encourage this is to communicate the company's policy clearly and directly. Here is how it might look in plain writing:

> *"If you are aware of serious misconduct such as (list examples) by anyone in the company, or associated with the company, report it immediately. There are a number of options for doing this:*
>
> - *The quickest and easiest way is to report it to your direct manager.*
> - *If you are concerned that your direct manager may be involved, or you are uncomfortable for any other reason, report it directly to a member of the Legal, Compliance, or HR teams, or to any senior manager.*
> - *If for any reason you do not wish to report the issue within the company, contact the following external agencies (list agencies)."*

One reason the sample text in this case is poorly written is that the writer started with the idea of establishing an all-encompassing 'policy' rather than focussing separately on the information needs of the key players in this scenario, viz the person reporting the misconduct and the managers tasked with following it up.

The extracted text was taken from an eight-page document that contained sub-headings *Purpose, Scope of Policy*, *Policy Statement*, *Confidentiality*, *Employee Rights*, *Untrue Allegations*, *Procedures for Making a Disclosure*, *Timescales*, *Investigating procedures*, *Communication with the Individual*, and finally *Speak Up Service Access*. Here are a few points that highlight what a confused mess this 'policy' document is:

- The three headings *Purpose*, *Scope of Policy*, and *Policy Statement* contain overlapping information. It would have made it much easier for the reader if the author had done away with these headings altogether and simply explained the policy in the document's opening paragraphs (which would be the Guidance Topic approach).
- Despite the document's title describing it as a 'policy', the author has included a section headed *Procedure for making a*

disclosure, which illustrates just how spurious it is to try to distinguish documents in this way.

- The grab-bag of section headings has no apparent logical structure, which means the reader must work hard to extract what is relevant to them and their circumstances.
- The document was published on the organisation's web site, which implies that its audience is the public at large. But the content refers specifically only to employees. Why was its publication not restricted to the organisation's intranet? In fact, is this not where an employee would expect to find this information if they were seeking it out? Putting on the public web site is actually likely to make it harder for the employee to find. Sadly, it looks like the document is intended more as 'window dressing' than a serious attempt to address an important corporate function. In other words, it is published on the public web site to give the public the impression that the organisation deals seriously with misconduct.

Plain writing is important and it is encouraged by abandoning the document pyramid and taking a Guidance Topic approach instead. Coincidentally, plain writing has received further impetus in the last few years by playing a key role in the application of Behavioural Insights or 'Nudge' approaches to change management.

Many Behavioural Insights initiatives involve the use of plain writing to positively influence behaviour by making communication clearer, easier to understand, more personalised and engaging, and more focused on getting a positive outcome for the intended recipients.

Sadly, in the terminology of Behavioural Insights, most internal policy, procedure, and instructional documentation today qualifies as 'Sludge' rather than 'Nudge' as the above examples illustrate.

2. Dynamically picturing the organisation

Leveraging the value chain

In the previous chapter we looked at how to transform policies and procedures from command-and-control instruments into knowledge sharing ones by conveying information in Guidance Topics. These act as atomic units of knowledge, each a minimum of self-contained useful information and collectively covering all team activities.

The question now is, how should Guidance Topics be characterised and organised into groups. Grouping is necessary for two reasons:

- so that they can be easily located
- to align them with the organisation's purpose

When a person looks for knowledge to support their work, they do it in response to their circumstances and the problems or difficulties facing them. This means that knowledge should respond to questions such as:

- What do I do if (a certain situation arises)?
- What do I do when (a certain event occurs)?
- How do I (complete a certain task)?
- How can I understand how (a certain set of activities inter-relate)?

Consequently, knowledge should be classified and presented in a way that provides direct answers to these sorts of questions.

The common characteristic of these knowledge-seeking questions is that they relate to people's activity. They relate to what they are trying to achieve in their jobs and by extension for the organisation.

The most useful basis then on which to classify knowledge is the way the organisation is structured to achieve its purpose – in other words, its business model.

This begs the question as to whether the organisation has clearly identified and articulated what its business model is. It needs to do this first before it can provide a useful context in which to develop and utilise knowledge. The essential theoretical groundwork for this task was prepared by Michael E Porter with his concept of a "value chain" (Porter 1985).

Porter and others (Magretta 2003) put forward the view that the purpose of an organisation is to create value, since this is the fundamental basis on which profit is earned.

This means that an organisation can be viewed as a set of primary activities that progressively add value in a linked "chain" that results in the outputs generated in terms of products and services.

In addition to these 'primary' activities, there are support activities that serve to facilitate them. In Porter's words

"The value chain displays total value and consists of value activities and margin. Value activities are the physically and technologically distinct activities a firm performs. These are the building blocks by which a firm creates a product valuable to its buyers."

Figure 2 is the schematic of the value chain drawn by Porter.

Figure 2 Porters 'value chain' schematic

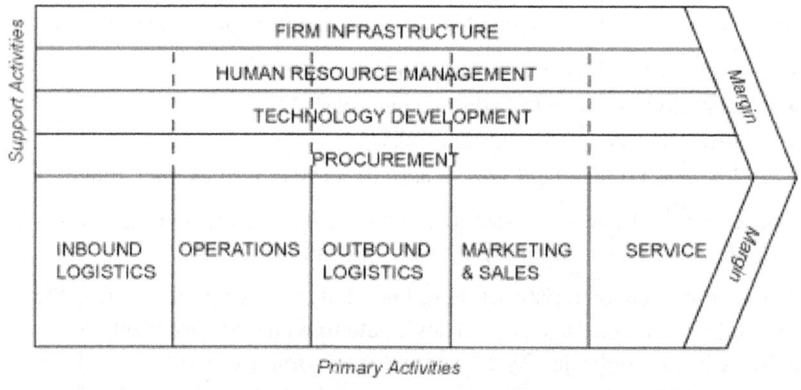

Porter's analysis was focused on the means by which organisations secure competitive advantage. We can apply the same concept to managing the knowledge that supports, maintains, and improves this competitive advantage.

By applying this value chain model as the conceptual context for developing and utilising organisational knowledge it is possible to link knowledge directly to the task of maximising value. This approach has the additional advantages that the value chain's zones of primary and support activity will:

- define the areas of the organisation that need to be covered by knowledge
- align with the way many organisations naturally divide their activity into specialised functions – this allows employees to locate relevant knowledge through their familiarity with its work context
- align broadly with the way management responsibility is allocated within an organisation – this makes it easier to identify who needs to take responsibility for developing the knowledge concerned

Porter's value chain therefore provides an effective overall context in which to develop and utilise knowledge. It provides the foundation for a dynamic taxonomy for organisational knowledge that ties directly to the organisation's purpose. As a component of Collective IQ, this is the framework in which 'atomic units of knowledge' can be both developed and deployed.

Developing a knowledge taxonomy for Collective IQ

This taxonomy needs a label and the best term to apply here is 'function'[3]. That is, each of the elements of Porter's value chain can be designated as a function.

Figure 3 redraws Porter's schematic based on this functional view of a generic organisation. It provides a picture of the organisation in terms of its unique activity structure, which is the source of its competitive advantage and at the same time identifies the areas of knowledge on which to focus Collective IQ.

---

[3] The Australian Government's AGLS Metadata Standard includes the term 'function', which is "used to indicate the role of the resource in terms of business functions or activities".

Figure 3 Functional view of the organisation

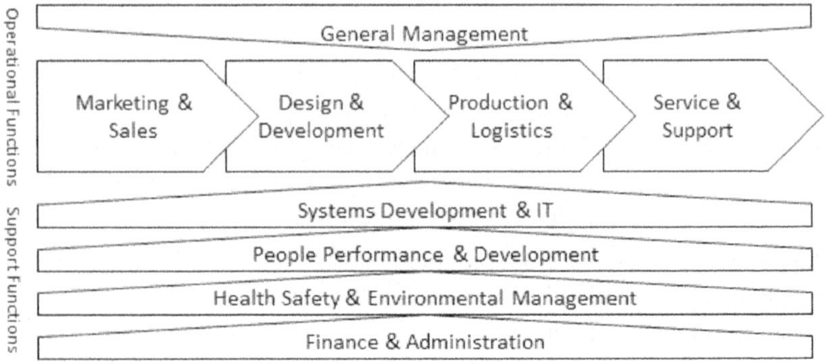

The approach in Figure 3 departs a little from Porter's view by substituting the classification 'General Management' for 'Firm Infrastructure' and grouping it with operational functions. The reason for this is that it provides a category of activity that controls and directs all other operational and support activities. This is an important additional concept to include when considering an appropriate structure for capturing and utilising organisational knowledge.

The General Management function is the 'kernel' of the enterprise. It determines what other functions are required and the general rules by which they will operate.

Having designated functions as the primary areas of specialised organisational activity, it makes sense to subdivide these further into useful component activities. The logical way to do this is to identify the *processes* that make up each function. Processes are effectively the value chains within functions.

This means that the terms 'function' and 'process' can be used to describe the same thing but at different levels. A function is simply a high-level process, and a process may just as effectively be termed a 'sub-function'.

This continuum is obvious when we identify the generic processes that constitute each function. Table 1 suggests what this division may look like.

In practice, the value chain, and therefore functions and processes, would be uniquely identified for each organisation.

Table 1 The value chain's functions comprise processes

Functions	**Processes**
General Management	▪ Governance and risk management ▪ Planning and reporting ▪ Managing quality and sustainability
Marketing & Sales	▪ Developing and implementing marketing strategies ▪ Managing customer relationships ▪ Processing and supporting sales transactions
Design & Development	▪ Developing new products ▪ Planning design activity ▪ Setting up new product production ▪ Controlling product specifications and modifications
Production & Logistics	▪ Managing supplier relationships ▪ Controlling production ▪ Storing and distributing products
Service & Support	▪ Responding to customer complaints and enquiries ▪ Providing repair and maintenance services ▪ Collecting and analysing customer feedback
Systems Development & IT	▪ Controlling and improving processes ▪ Managing information and controlling documents ▪ Providing office productivity support services ▪ Managing networks and data security

Functions	Processes
People Performance & Development	▪ Maintaining positive work environments ▪ Recruiting and inducting new workforce members ▪ Developing workforce capabilities ▪ Administering workforce support
Health Safety & Environment	▪ Protecting health safety and environment ▪ Dealing with accidents and incidents
Finance & Administration	▪ Maintaining financial control ▪ Managing customer credit ▪ Controlling and accounting for expenses ▪ Budgeting and financial reporting ▪ Taxation and statutory reporting

Functions and processes not only clearly map the territory that organisational knowledge needs to cover but, when used to classify that knowledge, provide an intuitively accessible means of locating relevant information when required.

Using the terms 'function' and 'process' to describe a two-level hierarchy for grouping activity in a value chain is somewhat arbitrary but there is justification for it in terms of how these words are broadly understood and applied in business[4].

Guidance topic definition and DITA

With this context in mind (that is, functions and processes that map the required knowledge territory) it is now possible to return to the idea of the

[4] For example, the AGLS metadata standard includes a 'Function' element, and the influential APQC in the US publishes a generic 'Process Classification Framework' that includes hierarchical categories for 'Process Group' (effectively a 'Function') and 'Process'.

Guidance Topic introduced in the previous chapter as a replacement for Policies and Procedures and create a more detailed definition for it.

As mentioned in that chapter, 'topic' is a documentation unit developed by the technical communication profession that is defined as "a title and content that is short enough to be specific to a single subject or answer a single question, but long enough to make sense on its own and be authored as a unit".

It is a key component of an open standard called DITA, which stands for Darwin Information Typing Architecture. DITA has the stated aim of "Advancing document creation and management specification that builds content reuse into the authoring process".

The Guidance Topic uses the DITA concept of a topic as a means of packaging business information in a more relevant and useful way by creating 'atomic' or 'modular' units of information. The key difference is that a Guidance Topic exists purely in the context of a particular organisation and is identifiable as a component of a business process that supports a function in the organisation's value chain.

The full definition of a Guidance Topic then, is:

> *"A title and content that is short enough to be specific to a single activity or circumstance, but long enough to make sense on its own and be authored as a unit, and which covers an aspect of a process that forms part of a function of a specific organisational entity."*

In other words, a Guidance Topic is a topic that will always have metadata associated with it that identifies the process, function, and organisation into which it fits.

3. Creating the right virtual environment

Communicating policies, procedures, and instructions in Guidance Topics makes the knowledge that they contain more relevant, accessible, and easy to assimilate.

Classifying them in a schema that mirrors the structure of the organisation as a value chain composed of inter-related functions and processes gives them a clear context and relevance to organisational activity. It creates a logical framework in which to develop and locate knowledge that is tied directly to the organisation's purpose, business model, and overall strategy.

To raise Collective IQ, however, requires a key third element and that is the capacity to document, organise, disseminate, and access knowledge quickly and easily. This is not possible with paper-based systems; it requires online access to computer networks, appropriate software, and the right approach to utilising them.

Characteristics of effective online systems

A base requirement is of course online, real-time access. In addition, effective systems require five key characteristics:

- file management
- efficient content access
- presentation flexibility
- distributed control
- linking and response mechanisms

These are listed in Table 2 below with details for each capability.

Table 2 Online system capabilities

No	Capabilities	Details
1	File management capabilities	read access, edit access, version control, metadata
2	Efficient content access	quick view and/or preview of file content (that is, not via download), persistent navigation interface
3	Presentation and location flexibility	flexible interface design, browse, search, filter
4	Distributed control mechanisms	direct authoring and publishing access for knowledge and authority holders
5	Linking and response mechanisms	hypertext links, sharing, feedback, discussion

These capabilities are present to a greater or lesser degree in the wide range of software options available on the market, but the problem is that it is often technically difficult to integrate them so that all characteristics can be easily utilised at the same time.

Product development has much to do with this because the relevant software products were first created for specialised purposes and then later enhanced to include some of the characteristics of the products they overlapped with. But not all, or in the same way.

Currently widely utilised software falls into five type categories, each with its own history and range of capabilities. These are:

- File Managers (FM) - This type is the ubiquitous application that is traditionally bundled with operating systems (the prime example being Windows File Explorer) that uses what may be termed 'Drive and Folder' classification. FMs have also blossomed more recently in the form of Cloud storage services (examples being Google Docs, Dropbox, and Microsoft's One Drive).
- Electronic Document Management Systems (EDMS) - This type developed from the need to manage large volumes of records in a more sophisticated way than FMs by using databases and metadata for control and classification. A number of these started life as

applications for indexing and tracking paper files (for example, TRIM and FileNet).

- Content Management Systems (CMS) or Enterprise Content Management Systems (ECMS) - These developed from web site authoring systems in order to meet the need to manage and control web pages and other files published online. These also employ databases and metadata for control and classification. An example is Oracle's Universal Content Management (UCM) system, which began life as the popular web site content management system Stellent, and Alfresco, which is an open-source alternative to proprietary systems of this type.
- Component Content Management Systems (CCMS) - This type is more recent and has been developed to provide a 'single sourcing' capability for multiple publishing modes. Examples are Ixiasoft and MadCap Flare. These systems have their genesis in specialised technical authoring software used to produce online help systems and technical and user manuals.
- Collaboration Systems or Groupware - This type was pioneered by Lotus Notes (now IBM Notes) but is now dominated by Microsoft's SharePoint. There are also specialised collaboration systems, such as Atlassian's IT-focussed Confluence, that include documentation capabilities in the form of Wiki-style libraries and reference file storage.

  Collaboration Systems /Groupware come closest to providing all the characteristics required to support effective knowledge management and Collective IQ.

All these types of systems, to varying degrees, are employed by organisations to manage the files that store their documented knowledge.

The least sophisticated and most limited of these systems is, ironically, the one that is most widely used to store and manage knowledge for internal use. This is the File Manager (FM), and the most commonly used of these within organisations is Windows File Explorer utilising a shared network location (a.k.a., shared network drive).

SharePoint: an example of a suitable system

To raise Collective IQ, organisations must, in addition to documenting policies, procedures, and instructions in Guidance Topics, move this knowledge from File Managers (shared network drives) into more sophisticated systems – ideally a Groupware System.

Because the most common of these is Microsoft SharePoint, I will use the latter to provide examples of how online documentation applications can be constructed to enhance Collective IQ. This is by taking advantage of the capabilities listed in Table 2 above.

SharePoint operates via most web browsers. It combines web site design capabilities with preconfigured, customisable databases for managing files (which are designated as 'Libraries') and for managing records (which are designated as 'Lists'). It also includes capabilities for automating interfaces between users and the contents of libraries and lists in the form of workflows.

For the moment, in terms of raising Collective IQ, we do not need to be concerned with SharePoint's 'list' and 'workflow' capabilities. Its web site design and 'library' capabilities are more than sufficient to begin with.

Figure 4 below shows what an application to support Collective IQ might look like in SharePoint. It is a single library that contains all the organisation's policy, procedure, and instructional information in the form of Guidance Topics.

These are classified by function and process. Figure 4 shows how the key elements that support Collective IQ are covered.

Figure 4 Interface to a SharePoint topic library

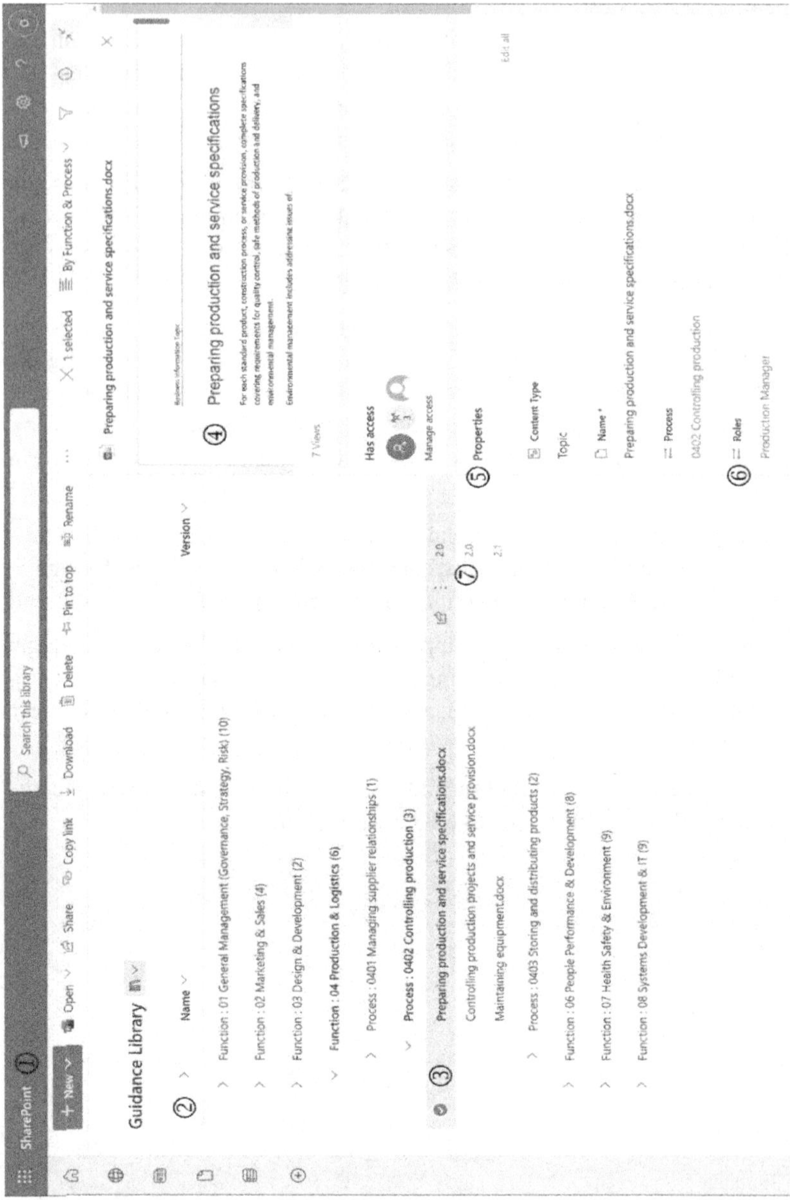

Here are explanations for each aspect labelled in Figure 4:

1. The library is accessed via a web page located in a SharePoint web site. Typically, this would be part of an organisation's intranet.
2. A 'table of contents' provides an overview of the whole library and a means of navigation. Topics are grouped under Function and Process headings. These can be expanded to browse the topics listed underneath, which makes best use of the screen space available.
3. Each topic has a meaningful title. When a topic title is clicked, it opens a view of the topic's content in its own browser window.
4. Selecting a topic (without opening it - indicated by a tick (3)) provides access to its information panel. This includes a preview of the topic's contents and metadata. All topics are in a standard format to provide a consistent user experience.
5. Each topic has associated metadata (referred to in SharePoint as 'properties'), which include function and process classifications that propagate automatically in the table of contents. Default metadata includes such information as the name of the person who last edited the topic, the original author, the dates of creation and modification, and so on.
6. A custom metadata element such as 'Roles' can be used to filter what the user sees according to their personal knowledge needs. Other custom metadata elements can be created as required.
7. Version numbers are assigned automatically. Both major and minor versions are tracked. Minor versions are drafts, which only authors can see. Major versions are the published versions that can be seen by all authorised viewers. The system also keeps a 'version history', with provision for comments on each set of changes. Access to groups of topics can be controlled so that only designated employees can edit them, which allows content management to be distributed and decentralised.

Those who are already familiar with SharePoint will recognise how the example has been created. It does not use any customised elements – everything is out-of-the-box. Those who are not familiar with SharePoint should regard the example simply as web pages presented to users that contain the interactive capabilities described.

Implementing an application such as this is the first major step towards raising the organisation's Collective IQ because it:

- assembles knowledge on how the organisation works (using the Guidance Topic form to document it and the Function/Process classification schema to scope and classify it)
- makes knowledge accessible to those who need it
- allows knowledge to be dynamically updated

The next chapter goes more deeply into the history of the development of the overall conceptual framework of Collective IQ before turning to more advanced practical issues.

4. Doug Engelbart and Collective IQ

Collective IQ is a concept developed by Doug Engelbart, the pioneering Research Engineer whose work on human-computer interaction led to the development of the computer mouse and Graphical User Interface (GUI).

He is most famous for presenting the “mother of all demos” in 1968 (O’Mara 2018), which was a demonstration of his team’s research work at the Fall Joint Computer Conference in San Francisco's Civic Auditorium. This demo paved the way for the world of interconnected personal computing that we now take for granted[5].

What has been largely overlooked since, however, is that these ground-breaking initiatives came about as a result of Engelbart’s overriding objective to 'augment the power of the human intellect'[6]. His concept was that by combining technological and human development (or co-evolution as he put it), it is possible to raise society's 'Collective IQ', which he defined as “the relative capability of a group of people with a common interest to cope with complex large-scale problems”.

To realise this concept, he and his team developed the pioneering IT software and hardware that was shown in the demo, and which has resulted in the enormous technical development that has taken place since: high speed networks are ubiquitous, use of multiple data processing and communication devices are taken for granted as part of work life, and

---

[5] Apple’s founders credit Engelbart with the invention of personal networked computing: “Engelbart! When you get to computers, Engelbart is great.” – Steve Wozniak, “Doug invented the mouse and the bitmap display” – Steve Jobs (Fisher 2018).

[6] Engelbart openly acknowledged that he was inspired by Vannever Bush's ideas published in 1945 on 'collective memory' and the technological storage and quick retrieval of information. Bush's ideas also strongly influenced the ideas of Ted Nelson, who coined the word 'hypertext', as well as Frederick Terman, the so-called Father of Silicon Valley.

applications aimed at virtual collaboration and knowledge sharing are cheap and freely available.

What have not been developed to the same extent, however, are the key concepts that led to the technology. These were identified by Engelbart (Engelbart 1982, 1990, 1995) as:

- Networked Improvement Community (NIC)
- Dynamic Knowledge Repository (DKR)
- Concurrent Development Integration and Application of Knowledge (CoDIAK)
- Open HyperDocument System (OpHyS)
- Bootstrap

Networked Improvement Community (NIC)

Engelbart's concept of a Networked Improvement Community or NIC is one in which a group of electronically connected people with a common purpose use technology cooperatively to tackle the complex problems that they jointly face. They work together to better understand their problem-space, to unearth the best candidate solutions, to decide how best to deploy operational resources and capabilities, to monitor progress, and to adapt effectively to unforeseen complications.

For organisations, the NIC concept is an evolution of the 'team'. It raises a team's profile from one in which the members are focused on cooperatively performing a fixed set of tasks to one in which they are also focused on sustaining and improving both how the team works and its outcomes.

Also, in the same way that teams interact with each other and form parts of larger teams (for example, manufacturing teams interact with design teams, and both are parts of the production team), NICs in organisations follow the same structure, with the difference being that their capabilities are augmented (Engelbart's term) by Collective IQ concepts and technology.

The notion that organisations must continually innovate and improve to stay competitive has been pervasive in management thinking since the so-called Quality Revolution of the 1980's and 1990's. In practice, however, the application of systematic improvement methodologies tends to be patchy.

The methodologies developed towards the end of the last century are now mostly promoted under the label 'lean', as in 'Lean Manufacturing' and 'Lean Construction' and are used very effectively by some organisations but are by no means universally applied. In many cases they are confined to parts of an organisation, often just those involving high-volume or large-scale production.

Approaches to innovation and improvement unfortunately still tend to be ad-hoc, episodic, or fragmented, so there is still great potential to engage employees (in general) and managers (in particular) in systematic improvement activity as part of day-to-day work (see Richford 2019). NICs provide the opportunity to realise this potential by creating a coherent and ongoing dynamic process centered on computer network-based tools.

Ultimately an Enterprise is potentially one big NIC that contains a network of numerous NICs based on specialised purposes and skills. The 'value chain' or 'function/process view' of the organisation explained in chapter 2 can be used to identify the groups that make up each NIC and to set up mechanisms for enabling NICs to interact with each other for the benefit of the enterprise as a whole.

Dynamic Knowledge Repository (DKR)

A Dynamic Knowledge Repository or DKR is an online location or 'container' in which knowledge is stored, utilised, and developed through direct user interaction. By 'knowledge' here is meant codified knowledge, that is, information held in discrete, human-viewable files that contain text, pictures, diagrams, illustrations, sound recordings, video recordings, etc., or some combination of these.

It is dynamic in the sense that its content is constantly updated by the members of a NIC (Networked Improvement Community). They ensure that at all times content is comprehensive, accurate, coherent, authorised, accessible in multiple ways according to purpose, and linked and marked-up to make it explorable in meaningful and useful ways.

CoDIAK

CoDIAK stands for Concurrent Development Integration and Application of Knowledge. It is the concept that you do not 'stop in order to learn'. The processes used to improve capability are integrated with, or even identical to, those used to support current action. This has the double benefit of fueling learning via immediate feedback, and quickly implementing newly devised improvements. DKRs, when properly implemented, are key mechanisms for integrating and applying new knowledge. They also provide the base on which new knowledge can be developed.

Open HyperDocument System

Open Hyperdocument System or OpHys is an ambitious technical challenge set by Doug Engelbart to create a new electronic document format and associated application systems for recording, sharing, retrieving, and linking information in a much more effective and collaborative way.

As a general rule, enterprises at the moment tend to manage the documents that contain their corporate knowledge in a fragmented and ad-hoc way. In addition, existing technologies, which are inexpensive and ubiquitous, tend to be underutilised. This is mainly because of a continuing paper bias in the way documents are stored and managed.

The most neglected technologies are hypertext and metadata. If an organisation sets out specifically to maximise its use of just these two technologies, it will begin to raise its Collective IQ significantly.

Bootstrap

Bootstrap or Bootstrapping is the notion that raising Collective IQ potentially operates on three levels, with the upper two levels aimed at boosting or accelerating learning capacity. On the first level (level A), raising Collective IQ increases the capacity to solve specific real-world problems and carry out complex tasks. For example, the problem or task might be how to design and produce a new model car.

The second level (level B) focuses on improving the methods and processes used to improve level A processes. To continue our example, this might involve looking at the way new model cars are designed and produced in general to come up with a better overall approach.

The third level (Level C) focuses on methods and processes used to improve Level B processes. In our example, this might involve improving the way manufactured goods in general, not just cars, are designed and produced.

These levels have implications for the nature of the NICs involved. For our examples above, level A would be addressed by a NIC comprising people working in a motor vehicle company's design and engineering function. Level B, on the other hand, would require a NIC with representatives from all the motor vehicle company's key functions (manufacturing, inventory, finance, sales, etc.). For Level C, a NIC involving people from multiple manufacturing companies, academia, and professional societies, not just motor vehicle manufacturers, would be required.

The latter also highlights one of the key aspects of Collective IQ in Engelbart's view and that is to address complex problems at a societal level that transcends the interests of individual organisations. In the case of our example, working together on improving general new model design and production methodologies benefits all manufacturing companies without impacting their relative competitiveness.

In Engelbart's words, the essence of bootstrap is "The better we get at getting better, the better and faster we'll get better."

Focusing on organisations

Engelbart aimed at society at large, but the aim of this book is more limited. Rather than focusing on society, the focus is narrowed to what can be done to raise the Collective IQ of enterprises. By 'enterprise' is meant any organisation that exists to provide products or services to society - whether it is in the private or public sector, and whether it is for profit or non-profit.

Why enterprises? When considered in terms of Engelbart's definition of Collective IQ, an enterprise is automatically 'a group of people with a common interest' and running an enterprise is also automatically a 'complex large-scale problem'. This means that enterprises are effectively tailor-made environments for the application of Collective IQ concepts.

Since enterprises are the productive heart of societies and are the almost daily pre-occupation of all its working citizens, the potential benefits are significant from a couple of points of view:

- The productivity and efficiency of enterprises will improve through better decision-making and smarter ways of working.
- People working in enterprises will learn ways of working together in computer-aided environments that they can apply to other contexts, such as professional networks, volunteer associations, non-work associations formed to pursue special interests, and political movements.

Once Collective IQ skills are acquired within an enterprise, they can be applied to second-order complex problems, such as how whole industries can work better by collectively solving shared (non-competition-related) issues, and third-order complex problems such as how to collectively improve improvement methods themselves.

If an effective model can be implemented for raising the Collective IQ of enterprises, then the benefits will spread to society at large.

5. Include all organisational knowledge in Collective IQ

Deconstructing organisational knowledge by transforming policies, procedures, and instructions into Guidance Topics and making them available in a well-structured knowledge repository is a major step forward in raising Collective IQ. It establishes a cornerstone for constructing an integrated and comprehensive digital structure for housing documented organisational knowledge.

Explaining how this is the case requires an overview of the structure of all an organisation's knowledge, and for this it makes sense to use the terminology usually employed when discussing knowledge management.

This overview is shown in Figure 5 below.

Figure 5 The structure of organisational knowledge

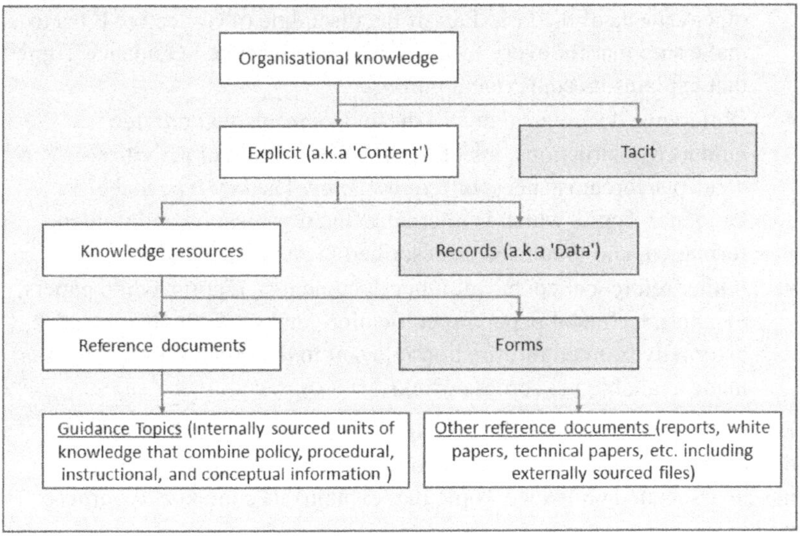

Figure 5 breaks down organisational knowledge in the following ways:

- All organisational knowledge is either explicit (documented) or tacit (in people's heads) (Nonaka 1995).
- Explicit knowledge is sometimes referred to as 'content' (as in web site 'content' for intranets - being the first widely used context in which digital information was aggregated for ease of user access). It can be split into 'knowledge resources' (these are files with information that convey meaning without the need to analyse or interpret) and 'records', which are raw data.
- Knowledge resources can be further split into 'Reference' documents and 'Forms'.
- Forms are a special category of knowledge resources. In this case 'forms' refer to standardised templates – not completed forms (which are records). Forms encapsulate knowledge by being integral to a task, usually to collect data to facilitate or record a transaction. They will often incorporate instructions for their own completion and even other aspects of the task. Part of the discipline of Collective IQ is to make sure that for every form, there is an associated Guidance Topic that explains its context and purpose.
- 'Reference documents' are all those documents that provide guidance, instructions, insights, or understanding of activities, concepts, circumstances, or requirements. The key type is the Guidance Topic, which is internal to the organisation and written, formatted, and organised as described in chapters 2 and 3
- 'Other reference' covers all other documents – reports, white papers, manuals, technical papers, specifications and so on. It also includes externally sourced information relevant to the organisation, which is made available via copies or links.

Part of the discipline of Collective IQ is to ensure that each of these documents is classified according to the process to which it relates and has an associated Guidance Topic that explains its context and purpose within the organisation.

In this overview, Guidance Topics are the principal or primary focus of knowledge resources because they are the first sources of written information accessed by employees when they need a better understanding of how things work or help to decide what to do in certain circumstances, or to check how to perform a particular task. These are all things where the accessibility, useability, accuracy, and authority of the information make a big difference.

Guidance topics function as the cornerstones of organisational knowledge because they are created within the context of the organisation's business model (as explained in chapter 3) and contain explanations of all the concepts, tasks, and relevant reference information that constitute the details of organisational processes.

Figure 6 below illustrates how this works. There are five key aspects:

1. Guidance topics contain all policy, procedural, instructional, and conceptual information packaged in decision and task focused units. These are classified by function and process.
2. Guidance topics contain references (preferably as hyperlinks) to all associated forms and other reference files.
3. Guidance topics also contain references (preferably as hyperlinks) to computer applications, or computer application instructions, where tasks have been automated or involve computer interaction.
4. Forms have a back reference to the Guidance Topic that provides information about where, when, and how the form is used. That is, how it fits into a process.
5. Other reference files have a cross-reference to the Guidance Topic that contains information about where, when, and how the file's information is relevant and therefore how it fits into a process.

Figure 6 The central role that Guidance Topics play in binding and managing knowledge resources

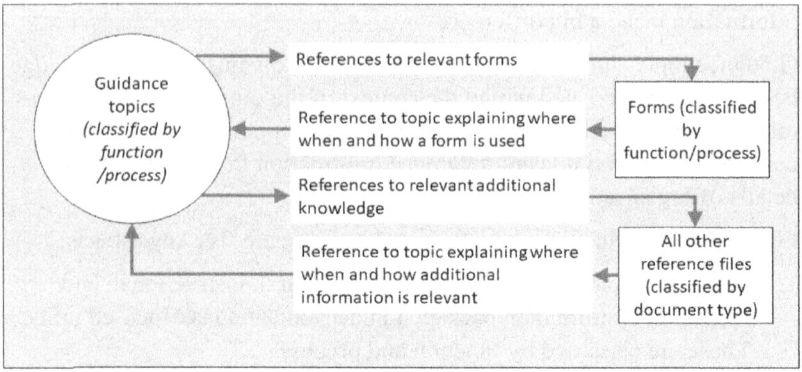

Knowledge resources are an organisation's memory

An organisation's knowledge resources are to Collective IQ what memory is to an individual. It is the store of knowledge that is both drawn upon when needed and expanded and improved by experience and learning. In individuals, memory structures itself automatically around a person's skills, interests, and desired capabilities. Organisational 'memory' on the other hand, requires overt structuring and management to be most effective. This is one of the key aspects of raising Collective IQ.

Typically, however, knowledge resources in organisations are left to accumulate in an ad-hoc way, without structure or rigor, which means that as 'memory' they are often weak and unreliable compared to their potential.

Adopting the approach illustrated in Figure 6 gives organisational memory the necessary structure and strength. The question remains however as to how best to implement it in a practical way.

Of course, it requires a networked digital environment and an application that supports Engelbart's concept of a Dynamic Knowledge Repository. The options available are outlined in detail in chapter 4 *Creating the right virtual environment* and the class of application that was the best fit was Groupware.

Groupware has several key capabilities in common. These are to:

- store files in discrete containers, usually called 'libraries' but sometimes called 'databases'.
- associate metadata (a.k.a. properties) with each file to support identification, classification, and searching
- create customised interfaces for users to interact with content, usually web-browser based
- control who can get access to the location and its contents

The first challenge therefore is to determine how best to 'package' knowledge resources in such a virtual environment. The simplest entry point for this for an organisation is to create an intranet site called 'knowledge resources' (or similar) and to populate it with libraries containing the documented information.

In the model being developed here, there will be a cornerstone library containing Guidance Topics, a second library for forms, and a third library for all other documents that comprise knowledge resources. This mirrors exactly the relationships set out in Figure 6.

The idea here is that each knowledge resource library is set up to retain a single document type as a primary way of identifying the nature of its contents, except that if it contains multiple document types then the primary means of classifying documents within the library is by document type. This provides a simple logical rule for scaling up a collection of knowledge resource libraries.

Apart from the Guidance Topic Library and the Forms Library all other document types go into the Reference Library unless there many documents of a particular type (for example, technical papers, legislation, computer user manuals, and so on), in which case it gets a library of its own.

There are three key things to note here:

- 'Document type' is an intuitive way of dividing up knowledge resources. It is also a key element in International and National standards for classifying information resources. This makes it a means of connecting an organisation's knowledge resources with the systems used by information management specialists in institutions such as web governance and standards bodies, public and university libraries, and government archivists.

- Guidance Topics and Forms are special document types in business contexts
- All other document types have relationships with Guidance Topics

Raising an organisation's Collective IQ requires a more sophisticated and comprehensive approach to the management of knowledge resources than is usually the case in organisations.

Document Type as 'package' for knowledge resources

While 'Type' is a recognised term amongst information management professionals, its use for classifying resources is not entirely straightforward.

The accepted International standard for classifying information resources is the DCMI Metadata Terms, known popularly as the 'Dublin Core'. In this standard, the element 'Type' is defined as "the nature or genre of the resource" and comprises a range of terms that includes 'Text'. Users of the standard are free to devise their own subdivision of this term, but the examples provided include "books, letters, dissertations, poems, newspapers, articles".

The Australian Government Locator Service (AGLS) is another Internationally highly regarded standard that is based on the Dublin Core. This standard goes a little further by including an element called 'Document Type', which has a suggested set of selections. These are listed in Table 3 below.

Table 3 AGLS Document Type vocabulary

- act
- agenda
- agreement
- blog
- checklist
- contract
- dataset
- diary
- digital certificate
- digital signature
- electronic message
- fact sheet
- form
- government gazette
- guidelines
- homepage
- index
- instruction (e.g. manual, handbook, tutorial, quiz)
- journal
- legislation
- letter
- log
- mandate
- media release
- meeting minutes
- memorandum
- metadata
- minute
- moving image
- note for file
- other
- policy statement
- procedure
- presentation
- promotion
- regulation
- report
- software
- sound
- standard
- still image
- wiki

Policy statement, procedure, and guidelines are highlighted in Table 3 because in the Collective IQ model these are regarded as types of information that are combined in a single document type, the Guidance Topic.

Guidance Topics and Forms are special document types

Guidance Topics and forms each get their own library, regardless of quantity. This is because they are the core of an organisation's knowledge.

The primary means of classifying their contents is by function and process, which supports Collective IQ in two ways.

First, it provides a logical means of classifying Guidance Topics in a way that makes information easy to locate. Guidance Topics can be written independently and then classified according to the process that they belong to (function classifications automatically follow processes because they have a parent/child relationship).

Second, the function/process schema delineates the areas of knowledge need within the business model and thereby helps to identify the Guidance Topics that need to be created.

Table 4 illustrates how a typical Guidance Topic library's contents are scoped using the function/process schema.

Table 4 Guidance Topic library typical function /process table of contents

Function	Process	Guidance Topics
General management and governance	Planning and reporting	▪ Overview of organisational planning ▪ Preparing and implementing plans and budgets ▪ Reviewing performance against plans ▪ Etc.
	Maintaining governance and risk control	▪ Overview of the governance framework ▪ Controlling and reviewing executive authorities ▪ Developing and reviewing policies ▪ Monitoring and controlling risks ▪ Carrying out risk assessments ▪ Etc.
Marketing and sales	Managing customer relationships	▪ Overview of customer relationship management ▪ Developing marketing programs ▪ Collecting and analysing customer feedback ▪ Etc.
	Processing sales	▪ Overview of sales activity ▪ Taking and following up orders ▪ Etc.
Production and logistics	Managing supplier relationships	▪ Overview of supplier management ▪ Controlling supply specifications ▪ Engaging suppliers and monitoring supplier performance ▪ Preparing and negotiating supply contracts ▪ Etc.

Function	Process	Guidance Topics
	Controlling production	▪ Overview of production management ▪ Controlling design specifications for production ▪ Monitoring production quality ▪ Etc.

Forms similarly benefit from being organised in this way. When placed in the context of the process to which they belong, their function is clearer, they are easier to find, and it is less likely that duplications will occur.

In addition, including metadata in the forms library that links a form to a Guidance Topic that explains where, when, and how it is applied, enforces a useful discipline that:

- ensures forms creators think carefully about how the form fits into a process and adds value
- provides staff and trainers with useful guidance and instruction
- makes automation easier by providing what is in effect a complete 'functional specification' for a new workflow or application

Guidance Topics relate to all other document types

Table 4 above illustrates how the Guidance Topic library is created to support the organisation's business model or value chain by developing knowledge around functions and processes. Covering the whole entity in a single library also provides the opportunity to place everyone's activities in an organisation-wide context that encourages knowledge sharing, cross-functional cooperation, and breaking down the 'silo mentality' that often restrains performance.

The Guidance Topic library's alignment with processes also enables it to anchor all knowledge resources in the operating context of the business. This can be achieved by storing all other knowledge resources in libraries classified by document type and by including metadata that links content back to the Guidance Topic library.

In summary, here are the key concepts behind structuring the management of an organisation's knowledge resources:

- Knowledge resources are stored in libraries according to document type.
- The primary document type is a Guidance Topic. The Guidance Topic library anchors knowledge resources to the organisation's business model (a.k.a. its 'value chain' expressed as an activity structure of functions and processes).
- A Form is an important document type that warrants a separate library. Each form links to a related Guidance Topic that explains its why, where, when, and how. Forms are precursors to automation.
- All other document types are stored in a single 'reference' library and grouped by type (using metadata). Each item or type links to a related Guidance Topic that explains its why, where, when, and how.
- If there are many items of a specific document type, they are stored in a separate library.
- Libraries proliferate by document type as the number of knowledge resources increases.

In the next chapter we will look at how to create Dynamic Knowledge Repositories that allow users to interact with organisational knowledge resources.

6. Building a Dynamic Knowledge Repository (DKR)

Building a DKR in SharePoint

How do you construct a virtual location in which to work with knowledge resources, or in Engelbart's vision, a Dynamic Knowledge Repository? What should it look like and how should it function? The conceptual foundations were established in the previous chapters so this chapter explores what a DKR might look like in practical terms in SharePoint.

SharePoint is based on web technology and its primary 'container' for content is a 'site'. A site is of course a web site, and it is created from a template that consists of a home page and several basic pre-loaded applications. For our purposes, the most important of these is the SharePoint library. Most site templates come with one library already created, which is named 'Shared Documents', but you can create as many libraries in a site as you wish.

A SharePoint library is a document repository that, for our purposes, has some important features:

- When a document is added to this library it is stored in a database and some key data are automatically collected and stored with it. These are its file name (with its extension so you know what software format it is in), an identification or ID number, the username of the person who added the document, the date it was added, the username of the person who last modified it, and the date on which it was modified. The system refers to these variously as 'properties' or 'columns' but they are, in fact, simply metadata.
- It also has an empty property called 'Title', in which you can enter a title for the document that is more descriptive than its filename.
- You may think the above unremarkable in that any File Manager will record similar information but there are some key differences:
  - The username is sourced from the groupware environment and its connection with your online identity within the organisation's network. It cannot be faked or tampered with (unlike the

'Author' property in an Office document) so it is, in effect, an electronic signature.
  - The identification (ID) number assigned to a document is unique to that library and similarly cannot be tampered with.
  - Properties can be used to list, sort, and filter documents much more flexibly than a File Manager.

- By selecting an option, the library can be made to store each version of a document as it is modified. Not only this, but it will also separate 'drafts' (identified as 'minor' versions) from 'published' versions (identified as 'major' versions). This allows you to discriminate between 'readers' and 'editors', in that readers see only published versions, leaving editors free to redraft documents until they are ready to publish an updated version. A 'version history' window can be opened to view the full update history of a document, complete with comments. Previous versions can also be backtracked or compared at will.
- You can define your own metadata. This is critical for us because of the importance to the DKR of Document Type and Function/Process classifications, which need to be added.
- The interface to a library, which is effectively a table of contents, is called a 'view'. When a library is created it has a single default view that looks much like a File Manager's list of drive or folder contents, but customised views can be added using selected metadata (both inbuilt and custom). Views can show groups of documents based on metadata (in which categories of contents can be expanded and collapsed on the screen), and the default view can be changed to a custom one.

These capabilities are important from the point of view of Collective IQ because:

- Author identity and document status are constantly automatically monitored, which means that the integrity of content can be tracked and managed. This is critical both for collaboration and for supporting content quality because it means that every document has provenance.
- Metadata can be added to identify the function/process to which a document belongs. Importantly this means that users are always able

to place the information in the context of the organisation's business model and overall operations.

The simplest way to set up a DKR for the organisation in SharePoint is to create a site called 'Knowledge Resources' and to populate it with three libraries. A cornerstone library containing Guidance Topics, a second library for Forms, and a third Reference library for all other documents that comprise knowledge resources. This mirrors exactly the relationships explained in previous chapters.

Using simple graphics for links, the home page of the Knowledge Resources site would look like Figure 7 (if this looks a bit plain, think of it as a wire frame to be rendered more attractive later by a graphic designer).

Figure 7 Knowledge Resources site home page

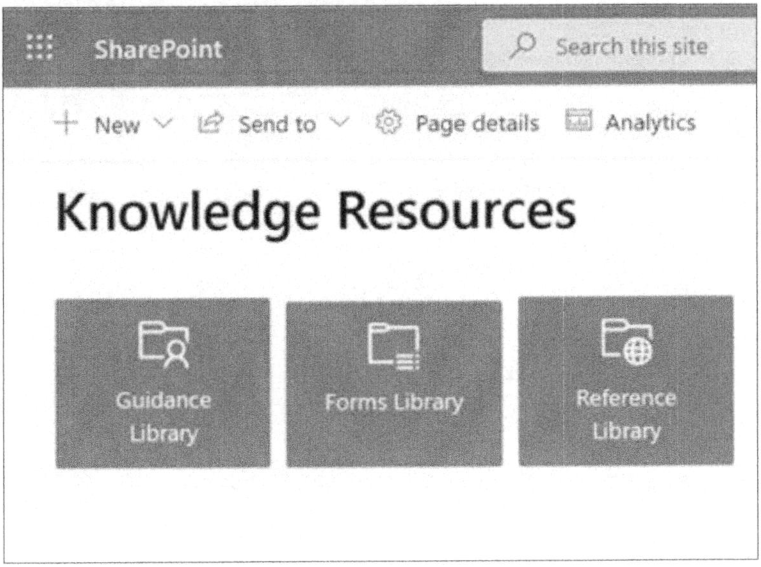

The tiles are links to the libraries. These links are replicated on the left (not shown) in a 'quick launch' column, which is part of SharePoint's default site layout. These are handy because they persist as you navigate to other pages or applications in the site.

A click on the Guidance Topic Library link opens the library in its default view Figure 8. This is an easily generated custom view that, using metadata, shows the organisation's functions as headings. Clicking a chevron (>) reveals the function's processes. And a further click reveals the Guidance Topics for that process, as shown in Figure 9.

Figure 8 Browsing the Guidance Topic library - function headings

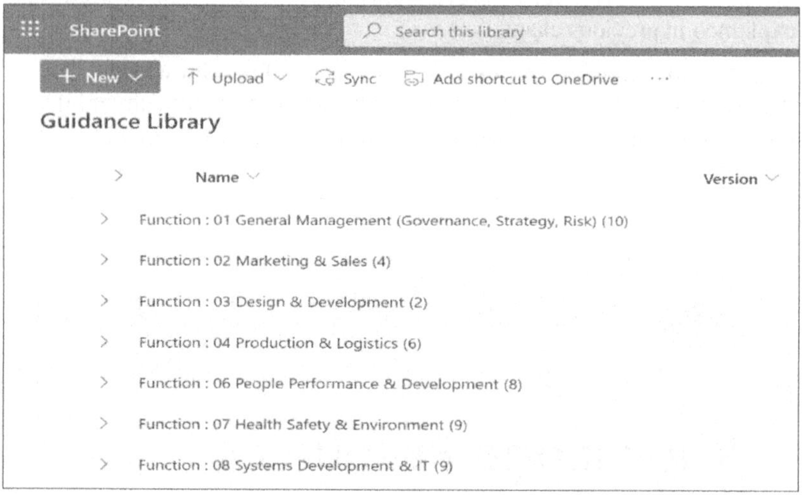

Figure 9 Browsing the Guidance Topic Library - processes and topics

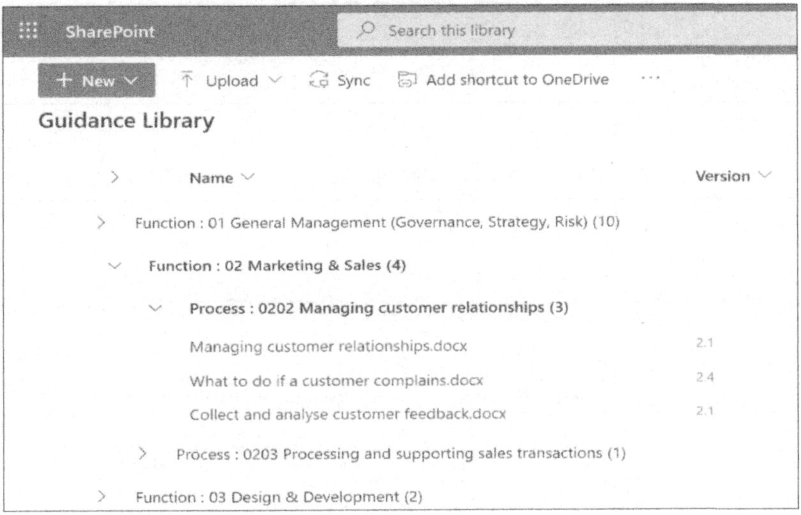

The SharePoint groupware environment is employed here purely as an example. Any other software that supports the same result is equally as good. The most important things about the above-described virtual environment are that:

- it provides a visual representation of the organisation's structure in terms of a hierarchy of activities (the function/process breakdown)
- it terminates in an encapsulated group of bite-sized pieces of information, viz 'atomic units of knowledge' in the form of Guidance Topics

A click on a Guidance Topic title opens it immediately on the screen so that its information can be quickly read and applied. The primary objective of a topic's content is to directly support efficient and effective task completion, optimal decision-making, and clear conceptual understanding.

Figure 10 shows what a Guidance Topic looks like on the screen in this sample SharePoint environment.

Figure 10 Sample Guidance Topic as it appears on screen

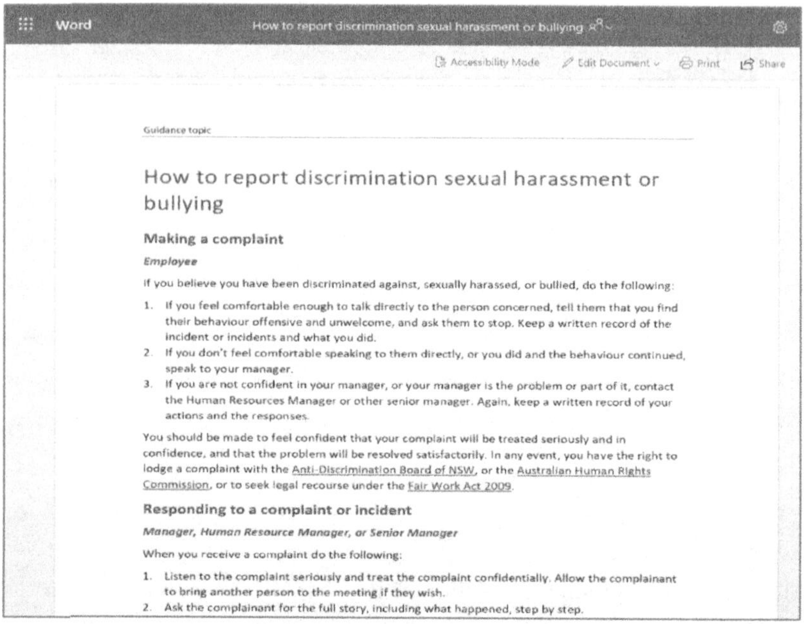

Word How to report discrimination sexual harassment or bullying

Accessibility Mode Edit Document Print Share

Guidance topic

How to report discrimination sexual harassment or bullying

Making a complaint

Employee

If you believe you have been discriminated against, sexually harassed, or bullied, do the following:

1. If you feel comfortable enough to talk directly to the person concerned, tell them that you find their behaviour offensive and unwelcome, and ask them to stop. Keep a written record of the incident or incidents and what you did.
2. If you don't feel comfortable speaking to them directly, or you did and the behaviour continued, speak to your manager.
3. If you are not confident in your manager, or your manager is the problem or part of it, contact the Human Resources Manager or other senior manager. Again, keep a written record of your actions and the responses.

You should be made to feel confident that your complaint will be treated seriously and in confidence, and that the problem will be resolved satisfactorily. In any event, you have the right to lodge a complaint with the Anti-Discrimination Board of NSW, or the Australian Human Rights Commission, or to seek legal recourse under the Fair Work Act 2009.

Responding to a complaint or incident

Manager, Human Resource Manager, or Senior Manager

When you receive a complaint do the following:

1. Listen to the complaint seriously and treat the complaint confidentially. Allow the complainant to bring another person to the meeting if they wish.
2. Ask the complainant for the full story, including what happened, step by step.

The sample in Figure 10 was chosen because it applies to all organisations. It also follows the writing guidelines in *Appendix I: How to write* a Guidance Topic in that:

- the title clearly describes its content
- it is written in the active voice to engage the reader
- it uses 'playscript' (Matthies 1961) to clearly identify who performs each step
- it uses hypertext to connect with other relevant information

What is meant by 'dynamic'

What makes the site dynamic is that it is being updated and added to at the same time as content is being accessed as a source of knowledge. While readers consult the topics they need, editors are revising, updating, and adding topics as processes are adapted, changed, and improved.

With its orientation towards knowledge sharing, as opposed to 'command-and-control', Guidance Topics are continuously revised to facilitate process improvements. Rather than being passive recipients of policies and procedures, employees are active participants in the application and development of knowledge resources.

This virtual environment also takes advantage of a SharePoint capability that gives flexibility to the way function/process metadata is deployed. In a dynamic organisation, processes are replaced and reorganised (under the impact of market forces, technological change, or strategic adaptation) and functions are changed as value-chains are restructured under the same forces. When this happens, Knowledge Resources libraries can be changed quickly by using 'look up' tables to classify topics by function and process. Changes to process names automatically flow through to all libraries, bringing their parent function names with them (Figure 11).

Figure 11 Function | Process metadata automatically linked to all libraries

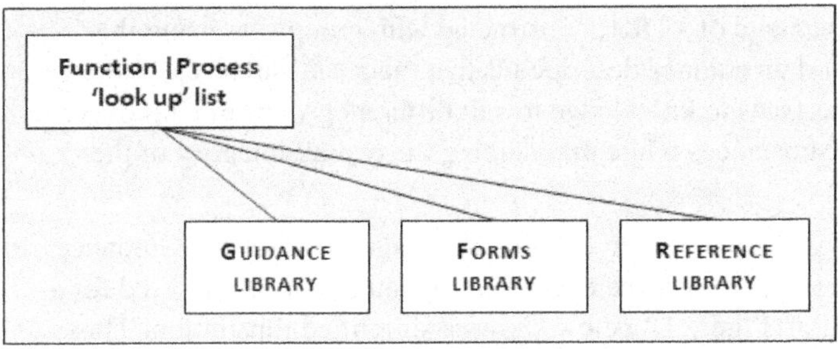

Working directly with the Function/Process list

Drawing function and process metadata from a separate list has another advantage. The list can be used to store other information about each process, including who is accountable for it and how its performance is monitored, measured, and improved.

This means that the list can be used as the focal point for the organisation's process improvement activity. As an organisational development tool, it sits comfortably with the use of a Process Classification Framework (PCF) as developed by the American Productivity & Quality Council (APQC).

The PCF is employed both conceptually and practically by many organisations around the world. The only difference with the way it is employed in the Collective IQ model is that is directly connected to the knowledge resources used to support day to day activity. This makes change management quicker and easier.

In practical terms it ensures that changed activity is well thought out, considered in context, and clearly communicated. In psychological terms it also makes a difference because changes are delivered into an environment of knowledge sharing that is acclimatized to continuous cooperative change and adaptation.

These ideas will be discussed in more detail in later chapters dealing with CoDIAK (Concurrent Development Integration and Application of Knowledge) and integration with continuous improvement.

Enriching the virtual environment

Another advantage of a DKR constructed with groupware using the principles and techniques described is that metadata can be used to create multiple interfaces to knowledge to suit different groups of employees in varying circumstances while maintaining the overall integrity of the system.

For example, it is possible to create a web page in which the Guidance Topics, forms, and reference documents are automatically filtered for a single function. Figure 12 below shows a simplified illustration. This alternative presentation allows those working in that area of activity to isolate the library contents most relevant to them without restricting their access to wider knowledge.

Figure 12 Filtered view of all library content for one function

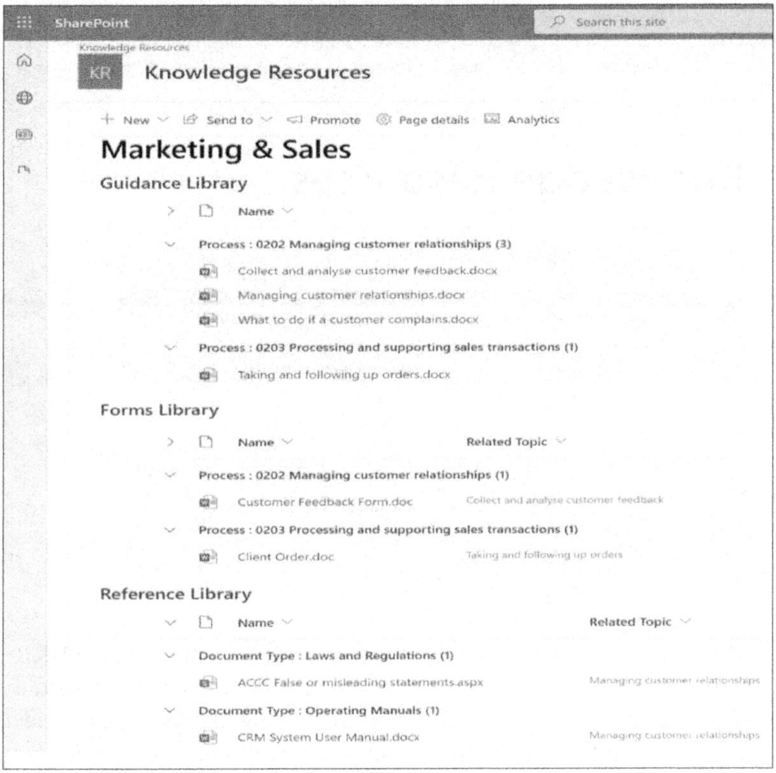

Creating 'landing pages' for each function and linking to them from tiles or icons on the knowledge resources site home page, is a simple way to provide intuitive high-level access to knowledge resources as shown below in Figure 13.

Figure 13 Business model representation provides intuitive access

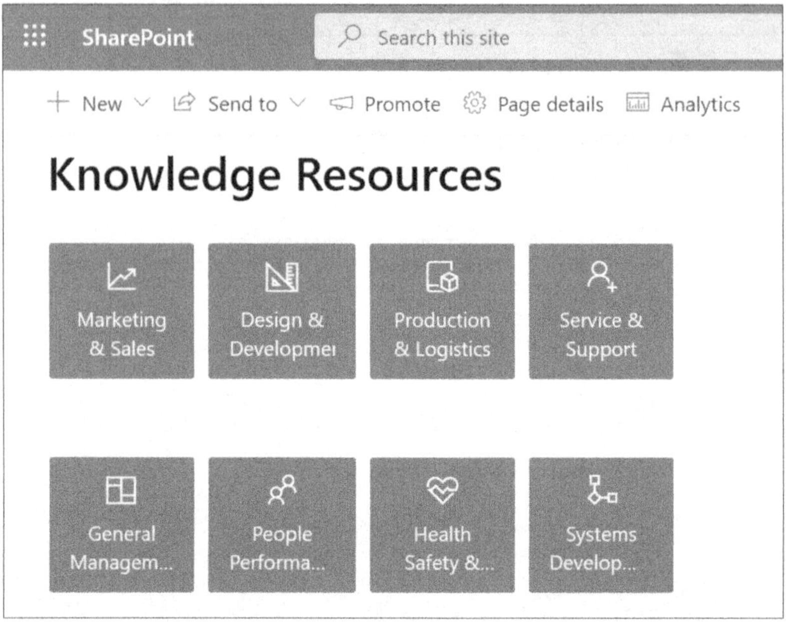

The Document Type and Function | Process model for structuring a DKR illustrated above gives everyone a clear framework in which to both draw on and contribute to organisational knowledge. And to do this in a context that encourages cross-organisational consistency, discourages duplication, and focuses content development on organisational objectives.

If you build it, will they come?

The evidence suggests that the answer is yes. If you make knowledge resources available to employees online, they will seek it out. And the better the quality, presentation, and coverage of that knowledge, the more enthusiastically it will be embraced.

In case studies carried out in two companies in the construction industry (van Ewyk 2016), high proportions of staff accessed online knowledge resources. For company 1, which had 250 employees, 82% of employees accessed knowledge resources during a 7-month data collection period and the average number of resources accessed per employee was 170.

Company 2, which was much larger with 9,900 employees, had a ratio of 50% and an average per employee access rate of 360 over a 12-month data collection period.

Although both companies went to significant lengths to deliver knowledge resources in accessible formats and in logically structured classifications, company 1 followed Collective IQ techniques much more closely. The effectiveness of these techniques was further confirmed by a case study carried out in an insurance company (van Ewyk 2018), which had approximately 100 employees. Data collected over an 18-month period showed that an average 15-20% of employees accessed knowledge resources each day. It needs to be born in mind, of course, that most insurance workers are office-based, and the work is more knowledge intensive than for many aspects of construction.

7. Implementing CoDIAK

CoDIAK and knowledge management

CoDIAK, for Concurrent Development Integration and Application of Knowledge, is Engelbart's concept of a virtual environment in which participants concurrently and collaboratively:

- draw on knowledge resources
- apply knowledge to achieve outcomes
- learn from those applications and make improvements
- deposit new knowledge based on what has been learned

The CoDIAK concept is a pivotal aspect of Collective IQ and it informed Engelbart's approach to the development of technologies to support it. These technologies are now substantially in place, but conceptual and practical barriers remain. The chapter *Including all organisational knowledge in Collective IQ* addressed one of the conceptual barriers by constructing a model of what constitutes organisational knowledge, viz:

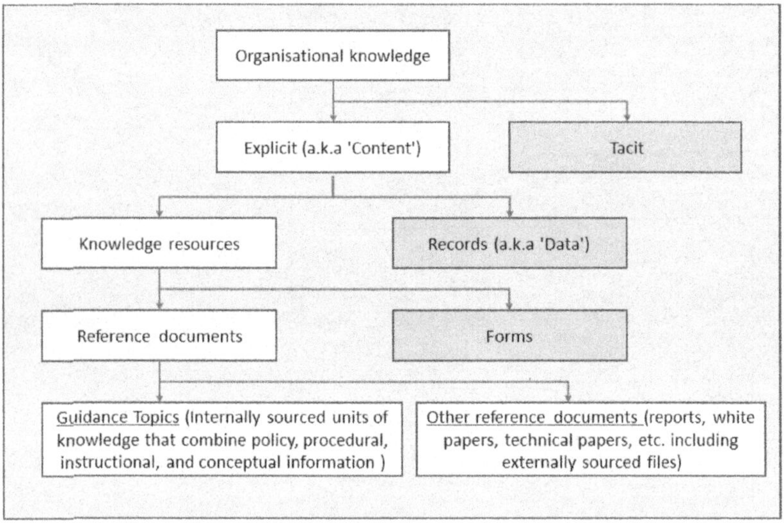

This model was used to identify and classify knowledge resources in order to design a practical Dynamic Knowledge Repository (DKR) as shown here:

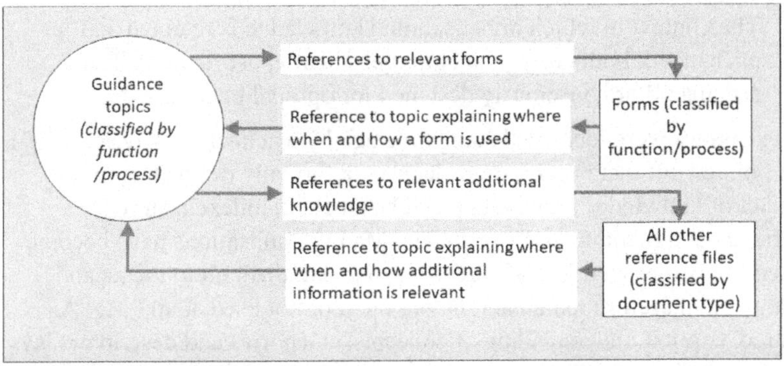

In this model, at the top of the hierarchy are the categories 'Explicit' and 'Tacit'. These are generally accepted terms to mark the distinction between personal knowledge (what's in our heads) and written knowledge (or more precisely 'codified' knowledge - that is, knowledge held in some objective form, which can be text, or a sound or video recording, and so on).

From the point of view of CoDIAK, however, the important thing is not the difference between tacit and explicit, but the relationship between them. In the words of one who coined these terms[7], "tacit and explicit forms of knowledge interact to create new knowledge ... (they) are not separate but 'mutually complementary' in that they dynamically interact with each other in creative activities by individuals and groups" (Nonaka and von Krogh 2009).

From a practical perspective, how does an organisation maximise this 'dynamic interaction' in the most effective way? There are two issues here:

- The important role played by explicit, codified knowledge is limited by a lack of clarity and consistency in the way it is formatted and presented.

---

[7] The tacit/explicit distinction was first coined by Ikujiro Nonaka, a pioneer of knowledge management, in his influential book "The knowledge creating company" co-authored with Hirotaka Takeuchi (Nonaka and Takeuchi 1995).

- The context in which organisational knowledge is retained and exchanged, both conceptually and physically, is often not clearly articulated or consciously designed for optimal benefit.

These issues were identified some time ago by Michael Zack (Zack 1999). He pointed out that organisations lacked "a formally defined, atomic packet of knowledge content that can be labelled, indexed, stored, retrieved and manipulated". And that "Many organisations have become so complex that their knowledge is fragmented, difficult to locate and share, and therefore redundant, inconsistent or not used at all." … "A shared, explicit understanding of concepts, categories and descriptors lays the foundation for effective communication and knowledge sharing in organisations." The latter he characterized as a 'Knowledge Architecture'.

The chapter titled *An atomic unit of knowledge* directly addresses Zack's concerns by introducing the concept of a Guidance Topic and the chapter titled *Including all organisational knowledge in Collective IQ* proposes the use of Document Type as a fundamental, metadata-standards linked, means of unit classification for all knowledge resources.

'Knowledge architecture', on the other hand, is a little more complex. In the field of knowledge management this term is used to refer to two things. It is used to describe what Zack is referring to – a structure or framework that makes knowledge easier to categorize, manage, and find – and, in addition, to the process of developing knowledge. The latter is also often referred to as the Knowledge Management Cycle (KMC), which typically contains the component activities of Identify/Create, Store, Share, Use, Learn and Improve (Evans 2014).

CoDIAK and the Knowledge Management Cycle are very similar, which connects Engelbart's notion of Collective IQ directly with knowledge management. The differences are that:

- CoDIAK emphasizes concurrent or simultaneous activity whereas the KMC is superficially a more linear activity
- CoDIAK is premised on the use of technology in the form of online tools whereas KMC is not
- KMC covers change and adaptation activity with 'learn and improve', while CoDIAK uses the words 'develop and integrate'

- KMC includes the upfront activities of 'Identify/Create, Store, Share' (in effect the creation of a knowledge architecture in Zack's terms) whereas CoDIAK settles for an all-encompassing 'develop'.

In the final analysis, the concepts of CoDIAK and KMC have essentially the same aims and components and, from the point of view of Collective IQ, can be combined into a single approach.

The other important aspect of a knowledge architecture is that it provides a structure or framework that makes knowledge easier to categorize, manage, and find. This is provided in Collective IQ by the proposed use of function/process classifications put forward in the chapter titled *Dynamically picturing the organisation*. This ties knowledge directly to the value chain/business model of the organisation and automatically provides a virtual space in which users can navigate intuitively. This is because the organisation's 'knowledge map' is also the topography of all its inter-related value-adding activities.

Practical application of CoDIAK

Collective IQ and knowledge management are a natural fit. The former brings technology, context, and discipline to the mix, while the latter brings a depth of commitment to learning from experience and valuing employee input and creativity.

Collective IQ gathers up typically fragmented and hide-bound approaches to policies, procedures, and other organisational information, and makes it more usable and accessible. It creates a conduit for knowledge management initiatives to make a direct impact on organisational activity and to facilitate improvements.

Outputs from knowledge management programs such as 'Communities of practice', 'Lessons learned de-briefs', and 'Case study analyses' can be fed directly into changes to operating methods via changes to Guidance Topics (or even process and function structures).

The immediate application of improvements or adaptations, in an employee population that accepts constant incremental change as normal, creates new knowledge that feeds the Knowledge Management Cycle.

On a day-to-day level, constant improvements and adaptations are supported by:

- Online feedback mechanisms – for example, links in each Guidance Topic that allow a user to give feedback directly on that link.
- Notification systems (emails, intranet news items, messages) that alert users to Guidance Topic changes.
- Internal social networking channels that routinely reference Guidance Topics to discuss changes and improvement.
- Open channels of communication to provide process improvement ideas
- Remits to process improvement project teams to include Guidance Topic changes in implementation programs.
- Availability of resources to assist with writing Guidance Topics and managing Dynamic Knowledge Repositories (DKRs).

Integration with formal training and learning activity

Training programs in an organisation often suffer from the same isolated and ad-hoc preparation as other knowledge resources. The materials prepared for training sessions often document policies, procedures, instructions, methods, and reference information, which at best duplicate what is already documented and at worst contradict it.

In Collective IQ, training and learning programs are prepared by first identifying and referencing the relevant Guidance Topics and then focusing on how to best provide training in terms of relevant instructions to trainers, sessions to run, and exercises and tests to complete.

There is a strong basis for arguing that all internal organisational training sessions are an opportunity to workshop relevant Guidance Topics and get valuable feedback from employees on the activities involved and how effectively they are documented.

8. Collective IQ and management

Collective IQ builds at each level of management

How does an organisation make Collective IQ happen? At an individual level, anyone tasked with writing helpful information can adopt the Guidance Topic approach (see the chapters titled *An atomic unit of knowledge* and *How to write a Guidance Topic*) and create more useful, readable, and accessible documents.

At a unit, department, section, or team level, managers can scope their team's knowledge needs by identifying their processes and specifying the Guidance Topics required to cover them. Most managers have access to some part of an online environment in which knowledge resources can be shared. If this is a groupware environment, then they can create a Dynamic Knowledge Repository (DKR) as described in the chapter titled *Building a Dynamic Knowledge Repository*.

The most beneficial level of implementation, however, is at an organisational level. This provides the opportunity for:

- a DKR to be constructed that fully reflects the value chain/business model
- the creation of tools and templates to support organisation-wide consistency in the way knowledge resources are prepared, presented, updated, and improved
- the creation of an environment in which all function and process relationships are evident and can be linked freely across divisional, departmental, or sectional boundaries

Organisational-level implementation, however, requires C-Suite commitment and participation. The CEO, COO, CFO, and CIO all need to sponsor, direct, and participate in the creation and maintenance of knowledge resources to secure maximum benefit. Constructing a comprehensive DKR and implementing CoDIAK for the whole organisation:

- recognises that knowledge resources are valuable assets that can be leveraged by being well maintained and improved
- promotes a common understanding of the organisation's dynamics

- allows anyone to drill down to a detailed activity level to check if values and strategies developed at the top level are being implemented
- enables knowledge resources to be linked to performance data
- connects high-level decisions with front-line activity

According to research conducted by McKinsey & Company, improvement is most effective when implemented organisation-wide (Sandoval 2019). According to this research: "improving the parts is much more likely to pay off if you first work to improve the whole, seeing it in its entirety from end to end. Then, the parts and the whole reinforce each other … organisations that break through silos in an end-to-end operating model operate more effectively."

Managers as custodians of knowledge resources

Employees have demonstrated that if good, accessible knowledge is available, they will seek it out and apply it. The challenge, however, is to develop and maintain knowledge resources in the quantity and quality required and this requires a change in managers' mindsets.

In management literature much is made of the importance of organisational knowledge but little is said about how the tangible embodiments of knowledge (knowledge resources) should be handled and by whom (this issue is explored fully in the paper 'Knowledge Assets and Knowledge Conversion' van Ewyk 2018).

Reference documents are often produced with the attitude that they are a sort of sporadic housekeeping chore. Their development is often prompted by an operational incident or crisis, or as a one-off adjunct to a change management project, or in response to external pressure in the form of legislation or regulations.

The Collective IQ approach, on the other hand, requires managers to recognise that the value to them of good reference documents is that they continuously communicate useful information to influence consistent desired outcomes. Outcomes that are directly connected to the organisation's goals. These outcomes include preventing incidents and crises from occurring in the first place and dealing with laws and regulations in a straightforward way that is integrated with the organisation's operations.

If they recognise this, they will treat reference documents as valuable assets that help them achieve their goals. As custodians of those assets, they are accountable for developing and maintaining them. And the higher the level of management, the broader the sweep of accountability and the greater the importance of a consistent, comprehensive, and timely approach.

9. Continuous improvement and automation

Integrating knowledge with continuous improvement

Continuous improvement, process improvement, Six Sigma, and lean management, are all terms for aspects of management practice that came out of the Total Quality Management movement that arose in the late 20th century, mainly driven in the West by the example of Japan's industrial successes, which were credited to the practices concerned.

Books written by W Edwards Deming ('Out of the crisis' 1982), Joseph Juran ('Quality Control Handbook' six editions from 1951 to 2010), Taiichi Ohno ('The Toyota Production System' 1988), Masaaki Imai ('Kaizen' 1986), and others, were highly influential in establishing these ideas.

These management practices are all in active use today under the banners of 'continuous improvement', 'process improvement' and 'lean'. They have these key aspects in common:

- Quality, expressed in terms of customer value and experience, is the organisation's primary focus. The essential idea is that the organisation's long-term financial returns, growth, and sustainability, all flow from this foundation.
- Customer relationships form the basis of operational processes and exist at every junction in the value chain, from the end user/consumer, through every linked organisational function, right through to the organisation's external suppliers of materials, components, and services.
- Processes determine the level of quality and how consistently it is achieved. People collaboratively control processes, and the aim is to minimise variation in performance outputs. (The term 'Six Sigma' is a nod to both the recognition of variation as a prerequisite to properly analysing performance, and to the use of sophisticated statistical process control methods – but the latter tends to be confined to high-volume manufacturing activity.)
- A culture of continuous incremental change to bring about improvements must be actively supported and promoted by managers in such a way that change, flexibility, and adaptation are a daily habit.

Collective IQ parallels, complements, and integrates with these key aspects in several ways.

The concept of a NIC (see Networked Improvement Community) is based on achieving quality outcomes and on the fundamental idea that those who work together on common goals will enhance their problem-solving abilities by sharing their knowledge online.

In organisations, NICs exist as a set of nested communities aligned with the organisation's functions (that is, the elements of its value chain as illustrated in *Figure 3 Functional view of the organisation* on page 20). Customer relationships are implicit in the value chain concept and therefore flow through to the function /process taxonomy used to structure the Guidance Topic library, which is the cornerstone of the organisation's knowledge resources.

When taken together to include all the relationships between value chain elements, the organisation constitutes a single combined NIC (as illustrated in Figure 14).

Figure 14 Nested organisational NICs

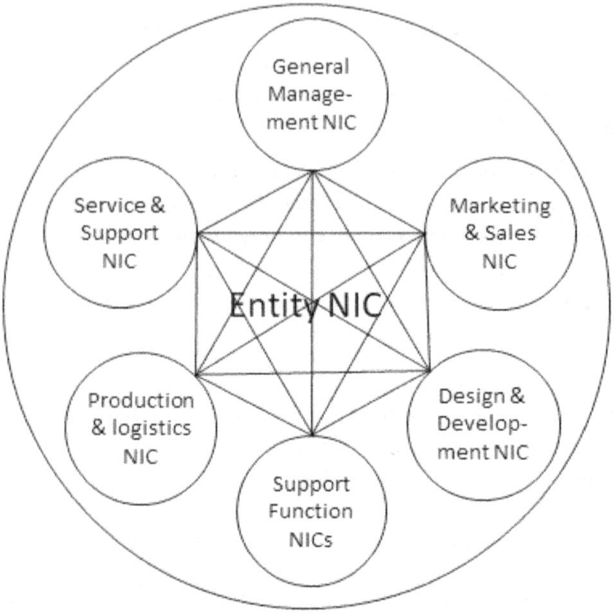

In effect, the NIC concept takes the idea of continuous improvement and adds the crucial dynamic of 'networked'. This links the powerful management ideas behind continuous improvement directly with an online communication environment and asks the question, "How do we best take advantage of online communication to support continuous process improvement?"

The response to this question is, of course, the implementation of DKRs and CoDIAK. Creating DKRs (see Dynamic Knowledge Repository) enhances continuous improvement by providing a 'knowledge currency' that is the basis of knowledge exchange within the NIC. It allows the NIC members to share a common detailed understanding of how processes work. In addition, this common understanding (or 'shared knowledge' or 'corporate memory') is structured and written to optimize its use by those who need to apply it. This means that it also functions as a direct support for daily activity and decision-making, for training, for induction, and for change management.

The driving idea behind the DKR is to inject speed, agility, and adaptability into the way policies, procedures, instructions, and other types of key organisational information are created, applied, updated, and enhanced. This is where CoDIAK comes into play to add to the means by which an organisation develops a culture of continuous change (both incremental and transformational). By linking improvement activity directly with knowledge development and application, it embeds continuous improvement solidly into the heart of organisational activity.

As a fit with contemporary ideas on continuous improvement, consider the following quote from an article entitled 'Continuous improvement – make good management every leader's daily habit' (Eichfeld et al 2017):

"Essentially, good daily management rests on a few basic disciplines: understand how people are actually delivering for customers, give people regular feedback and coaching, teach people how to solve problems, and create a physically and emotionally safe environment where people can engage in meaningful dialogue about their work."

These are the essential work environment dynamics that support continuous improvement. If you extend these to the networked work environment (via NICs), connect them with knowledge resources (via DKRs), and integrate feedback, coaching, and problem solving with knowledge resource development (via CoDIAK), then the power of continuous improvement activity will be both more far-reaching and sustainable.

Including knowledge resources (organised according to Collective IQ principles), also combats the 'procedural rigidity' exerted by traditional 'policy and procedure' documents, which can actively work against continuous improvement initiatives taken by teams. The traditional 'command and control' approach to documentation often results in a sort hardening of corporate arteries that slows or chokes off innovation and adaptation.

From the same article, their research found that, "On the ground, these organisations (those that make continuous improvement a daily habit) look and feel different from their peers. Their people collaborate much more often, and much more effectively, because they know what they're working toward and why. Every person has a deep understanding of how their daily activities contribute to the overall goal or strategy. There's a confidence that people can rely on one another to ask questions and contribute ideas."

Collective IQ techniques carry these work attitudes into the digital space and provide them with support and leverage. as well as helping to project them into a larger cooperative environment than the immediate team.

Their findings also underlined the need to apply the NIC concept to the whole organisation: "For continuous improvement to take off, the vast majority of management must consciously work together, as one very large team". And the value of devoting more time to "solving problems, improving standard work, building skills", all of which are enhanced, supported, and leveraged by an effective DKR and CoDIAK.

Continuous adaptive automation

Organisations rely heavily on forms to control productive activity. They are used to collect information in a consistent way to enable transactions to be carried out efficiently and effectively. They also record the results of transactions (and interactions), which provides evidence of results and captures information that can be aggregated and analysed.

The template for a form also encapsulates knowledge in that it implicitly directs activity in specific ways to achieve a result. When someone designs a form, their aim is usually to standardise a set of tasks in order to improve how they are handled. It is often the first step taken towards improving outcomes.

Unfortunately, it may also be an attempt to impose controls where none are needed, or to collect information speculatively that proves to have less value than the cost of collecting it. Forms can also be 'sticky' in the sense that if they are not changed to adapt to new circumstances or make improvements when opportunities are recognised, they hold the organisation back.

This is the reason why the DKR model described in the chapter titled *Building a Dynamic Knowledge Repository* requires that all form templates be linked to an associated Guidance Topic that specifies, at a minimum, the context for its use and places it clearly inside the process to which it belongs. In this way, it is always embedded in the process and accountable for its contribution to process performance.

Forms improve processes. They turn the informal into the formal. They transfer data accurately from one person to another. They create records that can be used to avoid duplication, help future action, and provide data for analysis and learning. And they are the logical precursors to automation. In conjunction with the Guidance Topics that explain when, where, and how they are used, they provide functional specifications for automation.

And tools are now readily accessible to non-programmers to immediately automate tasks using workflow applications such as Microsoft's Power Automate and PowerApps. This creates an exciting new dimension for continuous improvement in the form of what may be termed 'continuous adaptive automation', that is, the ability to automate tasks quickly and ad-hoc. Unfortunately, without a clear framework and discipline such as that provided by the DKR model, this sort of automation can quickly become messy, ill-directed, and counterproductive.

Following the DKR model also means that when a whole process or function needs to be rethought, restructured, or re-engineered, the DKR provides:

- complete and up-to-date knowledge of existing activity to inform business and systems analysts
- confidence that what is included is currently followed consistently (because it is supported by CoDIAK)
- a knowledge base that can be adapted to the newly restructured/engineered process (by retaining unchanged aspects – no babies thrown out with the bathwater)

Figure 15 Continuous improvement and adaptive automation illustrates how Collective IQ supports continuous improvement and adaptive automation.

Figure 15 Continuous improvement and adaptive automation

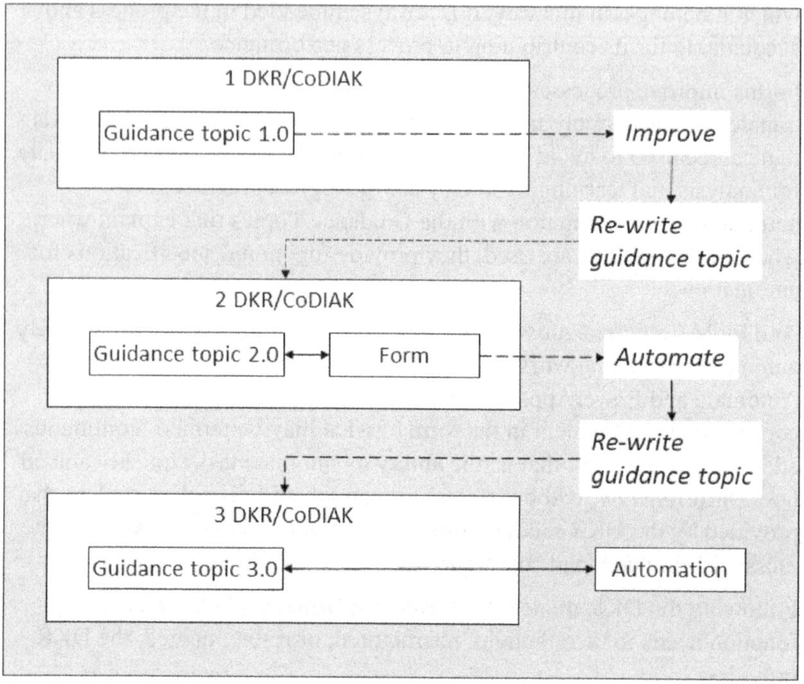

Figure 15 illustrates how the improvement process evolves in a DKR/CoDIAK environment:

1. A guidance topic is written to codify a task. It is used for training, it is made available for day-to-day reference, and it rewritten when improvements are made.
2. As part of an improvement initiative, a form is introduced to help facilitate the task. This standardises data capture and provides better control of the workflow resulting in fewer errors. The Guidance Topic is updated to cover use of the form. In conjunction with the change, the Guidance Topic is used to support retraining.
3. Based on the Guidance Topic and associated manual form, a further improvement initiative is implemented which replaces the form with online data capture and an automated workflow. The Guidance Topic is updated to include the user's interaction with the automated

system, and in conjunction with the change, the Guidance Topic is used to support retraining.

Compatibility with Agile

There are two management movements that have taken hold this century that are complemented by the concepts and techniques of Collective IQ. Both use the label 'Agile' in order to distinguish their precepts from traditional approaches, which are variously labelled 'Mechanistic', 'Bureaucratic', 'Trapped', and 'Hierarchical'.

The first is Agile Project Management, which is mainly applied to systems development. This movement began around the turn of the century and has been promoted under different methodological banners, including SCRUM[8] and DSDM[9].

The other use of 'Agile' is an approach to structuring an organisation to enable rapid adaptation and innovation while maintaining stable and reliable core activities. It has been promoted and developed by McKinsey & Co for the last few years. In their words (Aghina et al 2015):

> *"An agile organisation (designed for both stability and dynamism) is a network of teams within a people-centred culture that operates in rapid learning and fast decision cycles which are enabled by technology, and that is guided by a powerful common purpose to co-create value for all stakeholders.* ***Such an agile operating model has the ability to quickly and efficiently reconfigure strategy, structure, processes, people, and technology*** (my emphasis) *toward value-creating and value-protecting opportunities."*

---

[8] This term reputedly came from a rugby game analogy used to explain a novel approach to new product development which inspired the creation of this agile methodology (Schwaber and Beedle 2002).

[9] While the initials DSDM stand for Dynamic Systems Development Method, the methodology is intended for all types of projects (The Agile Business Consortium - agilebusiness.org).

Both these movements are aimed at overcoming rigidities inherited from past or traditional approaches to project management and organisational development. They focus on creating the ability to deliver value quickly and effectively by being connected and responsive to stakeholder needs, and by supporting a high-level of cooperation and collaboration. And the further this extends into the organisation without creating change-confusion, change-friction, or compromising the underlying imperatives of quality, safety, and environmental protection (i.e., stability), the better the organisation will perform.

In all these respects, Collective IQ (via NICs, DKRs, and CoDIAK) provides a highly supportive framework for achieving these aims.

10. The social framework and compliance

More than a bureaucratic exercise

One of the main drivers behind documenting policies, procedures, and instructions is to demonstrate to auditors and regulators that you have done your 'due diligence' and covered off their requirements. In some cases, legislation mandates it. For example, recent amendments to the Corporations Act (July 2019) require public companies to document whistle-blower policies, which cover the protection of whistle-blowers to encourage them to come forward.

While such requirements are rarely prescriptive in terms of how they are to be written, organisations mostly respond by writing documents that are implicitly aimed at auditors, often simply parroting the structure and content of the regulation. They are also invariably written in the passive voice using legalistic terms. This means that they are not written for the benefit of those who have to act on the information. The result is that the information is much less effective than it could be, and an opportunity is lost to integrate it with the organisation's operations and embed it in its culture.

Take, for example, the 'whistle-blower policies' requirement. Most organisations will take an omnibus approach and lump all information into a single, stand-alone, document called 'Whistle-blower Policy'. The Collective IQ approach, on the other hand, first asks the questions, "which organisational processes are involved?", "who are affected?", and "what do those affected need to know?".

In answer to the first question – the processes involved – a quick check of the Guidance Topic library identifies these as:

- Behaviour in the workplace (part of the 'People Performance and Development' function)
- Maintaining governance and managing risk (part of the 'General Management and Governance function)

In answer to the second question, there are three main groups of people affected

- all employees (those most likely to be whistle-blowers and who therefore will need protection)
- all senior managers and company officers (who are those authorised to receive whistle-blower disclosures and required to act on them)
- senior managers in governance and risk support roles (who are required to develop whistle-blower protection methods and monitor and report on their effectiveness)

This analysis also reveals any existing organisational activities in common with handling whistle-blower protected disclosures. Things like mechanisms for reporting wrongdoing or for conducting confidential internal investigations or for monitoring the effectiveness of policies and their implementation. If so, they can simply be modified to accommodate the whistle-blower protection legal requirements.

If not, new mechanisms developed can be most effectively documented in multiple Guidance Topics that cover the knowledge needs of each of the groups involved. The self-explanatory titles of the Guidance Topics would be as follows:

- "Whistle-blower protection – reporting serious wrongdoing" (aimed at employees and listed in the DKR under the process 'Behaviour in the workplace')
- "Whistle-blower protection – receiving and investigating disclosures" (aimed at senior managers and listed under the process 'Maintaining governance and managing risk')
- "Whistle-blower protection – monitoring and reporting on effectiveness" (aimed at risk management support roles and listed under the process 'Maintaining governance and managing risk')

Hyperlinks in the DKR create the appropriate connections between the Guidance Topics.

The net result is that the knowledge developed to comply with the law is now integrated into the organisation in a way that is easily located, accessed and understood by those who need it, and is therefore far more likely to be effective than the usual 'knee-jerk legal compliance' response. In addition, tracking compliance from management and auditing points of view is easily addressed by attaching the appropriate metadata to each

Guidance Topic. This enables compliance managers and auditors to isolate (by search or custom grouping) precisely where, how, and by whom, critical compliance requirements are to be handled.

These arguments apply to all compliance frameworks, whether legally mandated, self-imposed, or demanded by customers.

Three imperatives – quality, health, environment

There are three issues associated with organisational activity that are considered so important they are singled out for special consideration. These are quality, health and safety, and environmental protection. They are subject to many legislative and regulatory requirements, as are many other issues, but for organisations they are special because they are also covered by management system standards.

If an organisation implements these standards, they can be independently audited and certified as compliant. Once certified, they can claim that their methods for dealing with these issues are robust and reliable. In some industries, such as aerospace, construction, and engineering, customers give preference to purchasing from organisations that are certified and often make it a tendering condition. These standards are:

- ISO 9001 Quality management systems
- ISO 45001 Occupational health and safety management systems
- ISO 14000 Environmental management systems

There are over a million organisations world-wide[10] certified to the quality system standard alone so these standards are not a trivial part of organisational activity. Though the standards were developed along different timelines and under different pressures, they now exist as a sort of holy trinity of imperatives for how an organisation should operate in society.

These standards implicitly define an organisation's purpose in society as creating products and services that satisfy customers' needs (quality), without exposing customers, employees, or anyone else affected by their

[10] See www.iso.org/iso-9001-quality-management.html

activity to the risk of harm (health), and without creating pollution or degrading or diminishing the natural environment (environment).

It is difficult to argue that this is not a laudable prescription for organisational activity generally, and that none of these things should be subordinated to other issues. But of course, we know that this is not the case – organisations routinely subordinate them to short term financial or political considerations. The reasons for this are complex but there are two factors that contribute strongly to lessening the impact of these systems. The first is the tendency to design and document the systems in order to satisfy auditors rather than the needs of those working in the organisation, which makes them less effective for the reasons identified in the previous section *More than a bureaucratic exercise*.

The second is to treat the systems as ancillary to the organisation's core activities rather than integral with them (in many cases they are seen as simply a 'ticket to ride'). This also leads to the creation of separate internal bureaucracies for each standard. This happens frequently despite an entire section in each standard devoted to 'Leadership' with frequent references to 'Top Management', which results in these standards being implemented and driven by employees at lower levels of authority.

It is often the case that complying with one or more of these standards is the first impetus for the organisation to document its processes in a systematic and controlled way (which is an aspect of the standards). The result, in the absence of an existing structure that aligns with the value chain/business model of the organisation, is that information is structured to follow the clause headings in the standard. This can result in the standard's value not being realised efficiently or effectively in the organisation's operations, or even affecting them adversely because of the bureaucratic way they take effect.

Taking a Collective IQ approach mitigates these adverse factors in several ways.

- Documenting activities in Guidance Topics forces the requirements of the standards to be communicated in practical ways directly to those affected. This also happens within the operational context that make most sense to them.

- With a DKR (Dynamic Knowledge Repository) in place that already contains knowledge resources to support operational and support activities, compliance with standards is often simply a matter of minor changes to existing information (with change management already addressed by CoDIAK), or even simply indicating that what is already done is compliant.
- With Knowledge Resources already structured to mirror the organisational value chain (in terms of a bespoke function/process taxonomy, see the chapter titled *Dynamically picturing the organisation*), each compliance issue is integrated directly into the process to which it relates.

Implementing and maintaining these standards in a Collective IQ approach makes them more effective and easier to maintain. It can be summarized in this way.

1. Start with a DKR structured to support the organisation's value chain in terms of functions and processes. This means that knowledge resources are tuned to the needs of employees and actively managed by them (via CoDIAK).
2. To comply with a standard, look for where and how it impacts existing processes and modify them where it adds value.
3. Use the groupware's capabilities to 'map' compliance requirements to the relevant Guidance Topics. This can be done in a couple of ways:
   - The simplest approach is to use metadata to 'tag' each bit with the relevant compliance aspect (selected from a controlled vocabulary of terms) and then use this to search, filter, or group Guidance Topics by compliance requirement.
   - An alternative is to create a separate knowledge resource that lists each standard's requirements in the structure of the standard and then uses hypertext links to connect each of the standard's requirements to the relevant Guidance Topics. As a separate resource, it can be used to retain additional detailed information about each compliance requirement.

The third item above, particularly the second alternative, is the key to de-bureaucratizing the application of these systems. The compliance knowledge resource is only used by the organisation's compliance managers and external auditors to monitor compliance, provide consultative support to relevant operational employees, and facilitate both internal and external auditing. Any requirements in the standard that do not add value to the organisation can be explained in this resource and debated with the auditors.

By taking this approach, compliance is built into the organisation's mainstream processes – actual operational implementation of compliance is separated completely from compliance management. Figure 16 below illustrates how this relationship works online and Figure 17 shows what the resource might look like in SharePoint.

Figure 16 Create a compliance resource linked to Guidance Topics

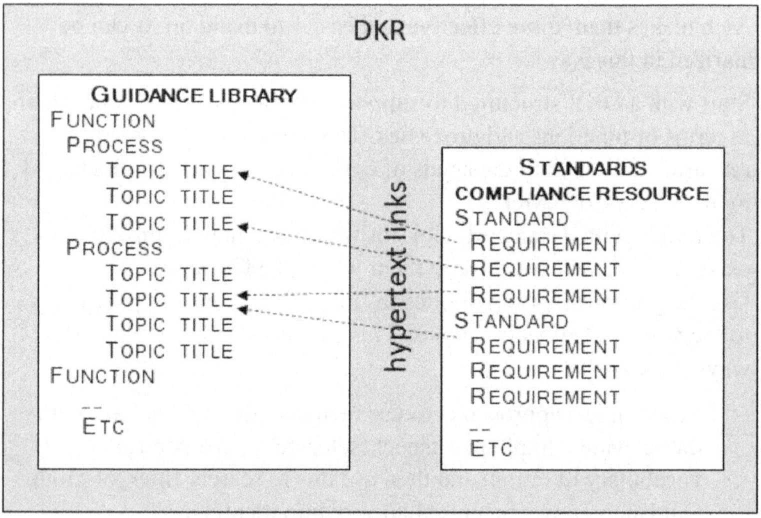

Figure 17 Compliance management resource in SharePoint

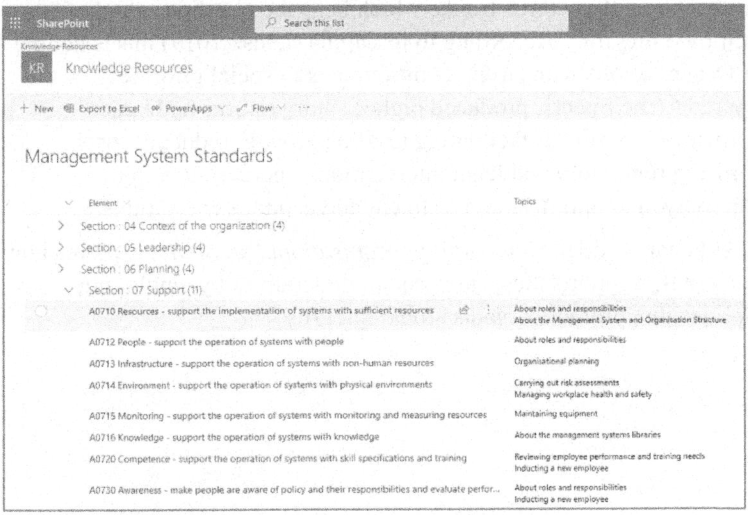

Managing compliance using groupware

Complying with auditable and certifiable standards to help manage quality, health, and environment, can be made to work much more effectively and easily when implemented using the Collective IQ approach described above.

But this approach does not need to be confined to these sorts of standards. It is valuable for all compliance frameworks. The advantages are the same whether they are externally audited or not. It treats compliance management as a sort of overlay on core organisational activity that asks and answers the question, "Where do we handle this issue in our processes?", which is the starting point for reviewing compliance, assessing effectiveness, and making improvements.

The social framework and profitability

Quality, health, and environment are society's three primary concerns in relation to organisational activity. This is exemplified by the importance placed by many organisations on compliance with the three management system standards that address them.

Despite this importance, many organisations choose to compromise performance on these three fronts, which begs the question - do they benefit by doing this? According to research (Henisz 2019) this is not the case. Organisations with strong commitments to social priorities, compared to their peers, produced higher equity returns and better performance in terms of facilitating top-line growth, reducing costs, minimising regulatory and legal interventions, increasing employee productivity, and optimising investment and capital expenditures.

The next chapter titled C*ombatting organisational stupidity*, explores how Collective IQ supports these three social imperatives by employing disciplines that promote organisational integrity.

11. Combatting organisational stupidity

Individual honesty versus organisational honesty

Is it smart for organisations to profit by making shoddy products?

Or by misrepresenting what their products and services will do, compromising the safety of employees, customers, or the general public, or doing environmental harm? Or by engaging in practices of dubious morality or legality on the assumption that they can spin, bluff, or manoeuvre their way out of blame or punishment?

The overwhelming majority of people would say no, it's not smart, it's stupid, because these behaviours are dishonest, immoral, illegal, or simply reprehensible, and ultimately lead to penalties, lawsuits, public censure, reputational damage, and even criminal prosecution. And yet, examples are legion of prominent and ostensibly respectable organisations doing all these things.

Well-publicized examples include the findings of the Australian Banking Royal Commission in 2018[11]. Banks charged fees for services that were not provided, advice fees were charged to the accounts of people who had died, loans were made to people without properly checking whether they had the capacity to repay, and financial advisers consistently acted against their customers interests in favour of their own, or even with outright dishonesty.

Further, the Commonwealth Bank breached anti-money laundering regulations 53,000 times[12], and the ANZ Bank manipulated a critical

---

[11] Banking royal commission calls for compensation, crackdowns and an overhaul of financial regulators, ABC News, 4 Feb 2019, https://www.abc.net.au/news/2019-02-04/banking-royal-commission-report-at-a-glance/10777188

[12] Commonwealth Bank faces shareholder class action, ABC News, 23 Aug 2017, http://www.abc.net.au/news/2017-08-23/commonwealth-bank-faces-shareholder-class-action/8833860

benchmark interest rate for its own advantage, for which it was sued by ASIC[13].

Examples from other industries include flagrant false or misleading advertising by large corporations. Optus was fined millions for misleading broadband ads[14], as was TPG[15].

Reckitt Benckiser, the large multinational that makes the popular painkiller Nurofen, marketed separate Nurofen-branded products for specific types of pain when they were the same, and charged up to double the price of the identically formulated general painkiller[16].

These things are wrong and yet they continue to occur. Is this because they are, in fact, the smart things to do to maximise profits in a hyper-competitive environment? Would a cost/benefit analysis show that the monetary gains made by these behaviours exceed the costs? If the organisation gets away with it, then the answer may be yes. If caught, then the answer may also be yes depending on the penalties and whether reputational damage can be overcome by increasing budgets for public relations and advertising.

It may be argued that these behaviours are simply organisations acting rationally by taking risks and suffering bad luck when the dice rolls against them. For example, it may be seen by some as smart to charge fees for no service when there is an opportunity to do so. It increases profit and if there is no scrutiny by regulators, or widespread publicity, then it is a

---

[13] ANZ sued by ASIC for interest rate rigging, The Sydney Morning Herald, 4 Mar 2016, https://www.smh.com.au/business/banking-and-finance/anz-sued-by-accc-for-interest-rate-rigging-20160304-gnaqg9.html

[14] Optus fined $5m for misleading ads, ABC PM, 23 Jul 2011, http://www.abc.net.au/pm/content/2011/s3263828.htm

[15] TPG fined $2m over misleading ads, ABC News 21 Jun 2012, http://www.abc.net.au/news/2012-06-20/tpg-fined-242m-over-misleading-ads/4082454

[16] Court finds Nurofen made misleading Specific Pain claims, ABC News, 14 Dec 2015, https://www.accc.gov.au/media-release/court-finds-nurofen-made-misleading-specific-pain-claims

successful strategy. If there had not been a Banking Royal Commission, the banks may have been able to spin away or frustrate the lesser pressure of individual complaints and regulatory enquiries. Even now, who is to say that the banks have not secured a net financial gain?

My suspicion is, however, that this is not the case. Penalties, lawsuits, reputational damage, and its impacts (future sales losses, demotivation of employees, reduced ability to attract talented new employees) exceeds the short-term gains.

What can be said for certain is that society suffers. Customers have been cheated, products or services have not delivered promised or expected value, employees and suppliers have been exploited, people have been harmed or placed in danger, and environmental harm has been done. While these behaviours may be smart in terms of financial gain, they are never smart in terms of the wider context of society.

When acting as individuals, the behaviours listed above are not widespread. We may cheat a little here and there in favour of our own interests, but generally we behave honestly to maintain our self-respect, and out of a genuine belief that it is good for society, and that what is good for society is good for us. If this were true of organisations, there would be far fewer instances of systemic bad behaviours. There may only be the actions of rogue employees to deal with but little more.

So why don't organisations act more intelligently? When critical decisions are made, why do they so often act stupidly and make bad choices? Why are there so many instances where the Collective IQ of an organisation is so much lower than an individual's IQ?

Organisational honesty and knowledge integrity

Part of the answer may lie in the way an organisation manages its knowledge resources. Taking Collective IQ seriously means maintaining the integrity of documented knowledge, and the integrity of the relationship between documented knowledge and those working in the organisation (directors, managers, supervisors, employees, and contractors).

If documented knowledge is treated as collective memory, its quality, and the degree to which organisational behaviour aligns with it, will determine the organisation's integrity. This significantly affects its ability to perform well for all its stakeholders (including society in general), and to sustain and adapt its performance in the future.

As an example of how the quality of documented knowledge affects alignment with behaviour, consider Reckitt Benckiser's misrepresentation of its Nurofen branded products. At the time that the behaviour occurred, like all large multinationals, Reckitt Benckiser, which now refers to itself as RB, had a 'Code of conduct'[17] that would have been similar in its coverage to the one that appeared on its web site in 2019. It would almost certainly have specified or implied the need for honesty and for compliance with the law. Despite this, the behaviour occurred.

Several arguments may be put forward as to why this was. It may be argued that there was a rogue employee acting against policy, or that the questions of ethics or legality involved were marginal, or that the executives concerned were unaware of the policy, or that the executives concerned simply did not take the Code of conduct seriously, or finally that those involved all thought it someone else's responsibility to apply the policy. Looking at each of these arguments in turn:

- It was the action of a 'rogue' employee. This is not likely because of the number of people that must have been involved and their seniority (product manager, marketing executives, package designers, production executives, and so on).
- The behaviour was considered ethically or legally marginal and open to interpretation. This is also not likely because the facts were clear cut – false claims were made and, on this basis, higher prices were charged for identical products.
- The executives concerned were unaware of the Code of conduct policy. This is unlikely too. Most organisations publish these things widely, include them in induction programs, and often insist on training sessions.
- The executives concerned did not take the Code of conduct seriously. The executives concerned may have decided that the potential profits

---

[17] http://www.rb.com/media/1509/sustainability-report-2011.pdf sustainability-report-2011.pdf

to be earned outweighed the risks of prosecution and censure. Combined with few internal barriers, this may have led to the action being taken.

- All those involved thought it someone else's responsibility. It may be that this sort of groupthink mentality existed, in which everyone involved thought it was somebody else's responsible to apply the Code of Conduct. As a result, no-one challenged the 'brilliant' marketing idea. This may have been compounded by the lack of a process and accountability for ensuring scientific accuracy in marketing activity.

I would speculate that the most likely cause lies someone in the last two arguments. The executives concerned did not take the Code of conduct seriously because, as an aspect of organisational knowledge, it had little value. It was seen, as is the case in so many organisations, as window dressing – a sop to regulators and the general public. No amount of internal publicity, or even training, changes this perception if the information concerned is not actively connected with other relevant information in the organisation aimed at guiding and coordinating behaviour. This is a failure of Collective IQ.

Documented knowledge as a fulcrum for action

What then should RB do differently? Not in fact what it did do, and that was to publish a barrage of documents on its web site pledging greater commitments to responsible and ethical behaviour.

At the time of writing (2019), RB published no less than five ethics-related documents on its web site. These were titled 'Responsibility' (a web page), 'Ethical business' (a web page), 'Business conduct' (a PDF file), 'Code of conduct' (a PDF file), and 'Ethical marketing' (web page). The latter contains the following text.

> *"Ethical business*
>
> *Honesty and transparency are fundamental operating principles at RB. We will ensure that all promotions and advertising is (sic) fair and honest and does not mislead consumers or is likely to do so by omitting material*

information OR presenting it in an unclear, unintelligible, ambiguous, or untimely manner.

All objective claims used in our communication and in our packaging & promotional materials must be backed by credible and documented evidence capable of withstanding scientific scrutiny. Where required, warranted disclaimers and qualifiers should be displayed at all touch points to ensure that the potential for miscommunication is minimized.

As a principle we believe we should 'go the extra mile' to ensure that the potential for misleading consumers is eliminated in our communication & marketing activities."

While the sentiments expressed here are laudable, there are many things about the information and the way it is provided that are problematic, and which point to its likely failure as organisational knowledge. The first is what may be called 'confused focus'. This is evident when the we ask the question, "who is talking to whom here?". The first paragraph appears to be a pledge to the world at large, principally consumers. The second paragraph, on the other hand, seems to be aimed at employees, insisting that 'scientific scrutiny' be applied to packaging and promotional materials. And the third paragraph returns the focus to the world at large to emphasise the pledge, albeit without adding any more meaningful detail.

As written communication, these statements can be immediately improved by applying Guidance Topic writing techniques in the following ways:

- Separating the text into two distinct pieces of information and making it clear who they are aimed at makes an immediate improvement. This means combining the first and last paragraphs to focus them on consumers and the public and giving them a meaningful title, viz:

  "Our ethical business commitment to consumers and the public

  Honesty and transparency are fundamental operating principles at RB. We will ensure that all promotions and advertising is (sic) fair and honest ... etc." "As a principle we believe ... etc."

This is a start to establishing some sense of credibility and integrity.

- Do the same for the middle paragraph to focus it clearly on the organisation's employees, viz:

  "To all staff members – ensuring fair and honest advertising

  All objective claims used in our communication and in our packaging & promotional materials must be backed by credible and documented evidence ...etc."

These simple changes make an immediate difference to the import and impact of the information, not only for the reader but importantly also for the persons in the organisation expected to act on it.

Unfortunately, the document has other serious problems. The content still lacks any convincing detail, there is no indication of its authority or who is accountable for it (is it from the Board, or the CEO, or the Marketing Director, or just the agency that prepares the web site?). Even with clarification of its focus, there is no indication how it will be effective in the context of the organisation.

RB operates in 60 countries, sells its products in nearly 200 countries, and has over 40,000 employees. This is a context that needs to be acknowledged when communicating anything to do with behaviour. RB's corporate headquarters manages this massive operation through its relationship with the operating units that report to it, so any processes aimed at assuring ethical behaviour should realistically be directed at the managers of these, who in turn need to interpret requirements and apply them in detail to their operations. This requires that knowledge be managed seriously in a connected and well-controlled way.

None of this is evident in the way RB has handled this issue. Publishing five disparate documents covering essentially the same ground does not clearly communicate a credible intent. There is no indication that they are part of a clearly thought-out structure of communication and interaction, which would be evident if the organisation's knowledge is managed in a coherent way.

In terms of appearance, the documents are colourful brochures or 'one-off' publications. There is no indication that they are serious working documents connected to organisational processes, as the issues concerned demand. They have inconsistent formats, contain duplications, duplicated hypertext links, and even hypertext links that point to the same document that they are in. None of the web page documents contain even basic metadata such as their Publication Date, Version, Context, and Author/Authority to assure the reader of their integrity. Several of the PDF documents contain some of this information, but all are different. In short, the contents of the documents are a mishmash of general information presented in a way that gives no assurance as to its effectiveness in addressing the issues that they purport to address.

Despite the flood of words, none of these documents are likely to have any effect on behaviour for the same reason – they are perceived as window dressing. They are only loosely connected with the organisation because of poor focus, content confusion, context confusion, and the lack of an internal 'neural network' of connected and coherent documented knowledge.

How then can it be done more intelligently?

Consider how it would be handled in terms of Collective IQ. In this context:

- It would be taken for granted that consumers and other stakeholders expect the organisation to obey the law and act fairly and honestly. Consequently, there is no need to publish a Code of Conduct (or Business Ethics or Responsibility document) on the web site. Its contents are only relevant to those inside the organisation.
- At most, if there have been instances of illegal or unfair practices, the intelligent course is to be upfront about them and report on what substantive changes have been made to prevent reoccurrence.
- The broader issue of web site communication would be tackled by:
  - considering how best to manage web site content
  - documenting the process
  - include the document in a Dynamic Knowledge Repository for general management processes so that it can be updated and improved (based on future feedback, evaluation, and discussion)
- The head office organisation would maintain a corporate level DKR (Dynamic Knowledge Repository) that covers all activities associated

with its relationship with the operating units that report to it (activities such as governance, planning, reporting, etc.). This repository would contain commonly agreed standards and methods, and be constantly subject to feedback, review and update. It would use groupware to manage the repository professionally.

- Within this repository, instead of a multiple 'responsibility', 'ethics', 'conduct' documents, it would contain a single document that identifies and emphasizes the legal and ethical risks specific to RB's activities (consumer health care and hygiene products, operations across multiple jurisdictions, etc.).
- Operating units, in response, would adapt their processes to give effect to the policy, with such things as, for example:
  - Changing the employment process to incorporate specific expected behaviours in the terms of employment and in employment contracts.
  - Changing the marketing process to require that all marketing plans and marketing materials be formally signed off as accurate by a senior technical or scientific officer.

By working in conjunction with a dynamic, inter-connected system of organisational knowledge, as illustrated above, the organisation creates a system that employees can rely on, follow, adapt, assess, challenge, review, and improve, and thereby act in a collective way. This creates far greater assurance that the organisation will consistently act intelligently.

I have singled out RB here only as an example, but the criticisms apply equally to all the banks named in the Banking Royal Commission, to Optus, to TPG – in fact, to far too many large organisations.

12. Intelligent free enterprise

Engelbart's vision, Tim Berners-Lee, and the future

> *"I became motivated (committed) in 1951 to improving mankind's capability for dealing with its pressing problems, especially those over-taxing our collective capability to cope with complexity and urgency. I visualized people collaborating interactively on visual displays connected to a computer complex."*

Doug Engelbart used these words when addressing a conference in 1992 to recount how he began to work on the ideas that he grouped under the banner of 'Collective IQ'.

Engelbart openly acknowledged that he was inspired (as were many computer scientists of his generation) by an article by Vannevar Bush entitled "As We May Think" published in 1945. Bush was an eminent engineer, inventor, and science administrator. In his article he envisioned a machine, dubbed the 'Memex', that would augment a person's memory by storing all their documents and be "mechanized so that it may be consulted with exceeding speed and flexibility".

Engelbart expanded on Bush's concept of "augmented memory" to include enhanced collaboration, so that his vision became one of 'augmentation of the human intellect'. In 1962 he submitted a report to the Director of Information Sciences of the Air Force Office of Scientific Research entitled "Augmenting Human Intellect: A Conceptual Framework" detailing his ideas. Just six years later these ideas were realised in their nascent form and brought to the world's attention in the 'Mother of all demo's'. The technological impact of this event was seismic. In an article marking its 50th anniversary, Wired Magazine[18] described it in these words:

> *"Engelbart gave a demonstration of the "oN-Line System" at the Fall Joint Computer Conference in San Francisco on*

---

[18] "50 Years Later, We Still Don't Grasp the Mother of All Demos" Wired Magazine 12.09.2018, Klint Finley

Dec. 9, 1968. The oN-Line System was the first hypertext system, preceding the web by more than 20 years. But it was so much more than that. When Engelbart typed a word, it appeared simultaneously on his screen in San Francisco and on a terminal screen at the Stanford Research Institute in Menlo Park. When Engelbart moved his mouse, the cursor moved in both locations.

The demonstration was impressive not just because Engelbart showed off GoogleDocs-style collaboration decades before Google was founded. It was impressive because he and his team at SRI's Augmentation Research Center had to conceive of and create nearly every piece of technology they displayed, from the window-based graphical interface to the computer mouse. .... You can draw a line from the technologies introduced at the "Mother of All Demos," .... to the internet, the web, Wikipedia, the Macintosh, Microsoft Windows, Google Docs, and a host of other technologies."

The impact of the demo was huge in that it showcased capabilities that had never been seen before but Engelbart was disappointed that it did not have the intellectual impact that he hoped for.

From the same article:

"To Engelbart, his work was never about the technology itself, but about helping people work together to solve the world's biggest problems. He believed that computers would be an important part of enhancing human abilities, but he also believed technology needed to be part of a systematic approach to problem solving and collaboration. Engelbart believed people should focus on creating feedback loops to improve their own effectiveness explains Jeff Rulifson, the computer scientist who developed much of the software on display at the Mother of All Demos. "The idea was to create tools and then use those tools to improve the tools," Rulifson says. Instead of making the tool once, it would be continually improved, based on the experiences of its users. As the tools improve, they make it possible to

> *make new, more useful tools. Engelbart called the approach "bootstrapping," named for the bootstrap circuit in radar systems."*

'Bootstrapping' (Bootstrap) and 'Open HyperDocument System' (OHS) are the two aspects of Collective IQ that I have not addressed so far. Engelbart defined bootstrapping as a "strategy fostering networks of organisations collaborating on capability infrastructures, explicit co-evolution of tool-systems and human-systems (skills, knowledge, procedures)" (Engelbart 1995). And Open HyperDocument System he defined as "an integrated, seamless multi-vendor architecture in which knowledge workers share hyperdocuments on shared screens. Hyperdocuments are multimedia files supporting linking and multiple object types." (Engelbart 1995)

The reason for not including Bootstrap and OHS is bound up in the brief history noted above. Despite its extraordinary technological capabilities and leaving aside transaction processing and general communication (including social networking), networked computer environments rarely function as more than Vannevar Bush's 'Memex' device. That is, a means of enhancing personal memory. There are some examples on the internet of collaborative knowledge repositories, including GitHub and Wikipedia, but even Wikipedia is limited in that it is a 'knowledge accumulator' (albeit dynamically maintained and expanded), not a generator of new or original knowledge.

Within organisations, the same shortcomings exist that disappointed Engelbart. While there are many organisations that apply continuous improvement techniques and the people involved are connected via computer networks, there is no evidence that they utilise Dynamic Knowledge Repositories that are continuously maintained using CoDIAK techniques. On the contrary, such activities tend to be project-based and confined to units or sections within organisations, with any documented knowledge siloed and/or unintegrated with other organisational processes.

The reasons for this are those set out in the early chapters of this book – the lack of an atomic unit of knowledge, an outdated approach to documenting policy, procedure, and instructional information, the absence of an explicit organisational framework (value chain – function/process view of the organisation), and the lack of an organisation-wide knowledge sharing culture orientated towards problem solving. In short, organisations need to be proficient in NICs, DKRs, and CoDIAK before they can participate in Bootstrapping as envisaged by Engelbart.

Open HyperDocument System (OpHys) on the other hand, is a technical challenge that is yet to be overcome. Hypertext links, while integral to the operation of web-based information, are technically crude. They are manually inserted, are dependent on the technical requirements of the content management systems in which they reside, and often break when information is updated or moved. It is an issue that may be resolved via the 'semantic web' or other initiatives aimed at making and maintaining connections automatically.

When Engelbart developed his ideas, and he and his colleagues demonstrated them, they helped significantly to ignite the explosion in technological development that has resulted in the predominance of online interaction and transaction processing that exists today. This would not have occurred, however, without the creation of the World Wide Web by Tim Berners-Lee in 1989.

In 1996, in a paper reflecting on the history of the internet to that point (Berners-Lee 1996), he acknowledged Engelbart's and others' contributions and his own motivation as follows:

> *"The origins of the ideas on hypertext can be traced back to historic work such as Vannevar Bush's famous article "As We May Think" in Atlantic monthly in 1945 in which he proposed the "Memex" machine which would by a process of binary coding, photocells and instant photography, allow microfilms cross-references to be made and automatically followed. It continues with Doug Engelbart's "NLS" system, which used digital computers and provided hypertext email and documentation sharing, with Ted Nelson's coining of the word "hypertext". For all these visions, the real world in which the technologically rich field of High Energy Physics found itself in 1980 was one of incompatible networks, disk formats, data formats, and character encoding schemes, which made any attempt to transfer information between dislike systems a daunting and generally impractical task."*

In the same paper, Berners-Lee goes into some detail about the aims of the web:

> *"We have seen that the Web initially was designed to be a space within which people could work on an expression of their shared knowledge. This was seen as being a powerful tool, in that*
>
> - *when people combine to build a hypertext of their shared understanding, they have it at all times to refer to, to allay misunderstandings of one-time messages*
> - *when new people join a team, they have all the legacy of decisions and hopefully reasons available for their inspection*
> - *when people leave a team, their work is captured and integrated already, a "debriefing" not being necessary*
> - *with all the workings of a project on the web, machine analysis of the organisation becomes very enticing, perhaps allowing us to draw conclusions about management and reorganisation which an individual person would find hard to elucidate*
>
> *The intention was that the Web should be used as a personal information system, as a group tool at all scales from the team of two, to the world population deciding on ecological issues. An essential power of the system, as mentioned above, was the ability to move and link information between these layers, bringing the links between them into clear focus, and helping maintain consistency when the layers are blurred."*

The first three dot points above precisely describe the contents of an effective Dynamic Knowledge Repository (DKR) but more than 20 years later, how many organisations can honestly claim to have such resources? The last paragraph above describes a 'Bootstrap' environment, which again is far from being achieved.

In the same paper, he goes on to say:

> *"At the time of writing, the most famous aspect of the web is the corporate site which addresses the general consumer*

> *population. Increasingly, the power of the web within an organisation is being appreciated, under the buzzword of the "Intranet".*

A novelty in 1996, the Intranet is a standard feature of most organisations now, certainly all large ones, but the extent to which they are effective as "a group tool at all scales from the team of two (upwards)" is arguable. Certainly not 'from the team of two to the organisation as a whole'. The argument of this book is that the barriers to this are conceptual and cultural, not technical. Adopting the Collective IQ approach set out in this book will go a long way towards overcoming these barriers and realising these visions of enhanced problem-solving.

In adopting Engelbart's concepts and components of Collective IQ, I have also sought to integrate them with:

- development of more effective means of written communication in online environments using DITA concepts
- the practical application of groupware systems
- knowledge management in terms of metadata standards and the development of online resource management science
- process development frameworks and support for continuous improvement activity

With these things in mind, the ultimate aim is to provide a practical path for organisations to develop 'a group tool at all scales from a team of two to the organisation as a whole'. Once there, perhaps they will aspire to collaborate with other organisations to create "networks … collaborating on capability infrastructures" and even "a group tool to operate for the world population deciding on ecological issues".

Capitalism redefined and Collective IQ

There is increasing pressure in western society to re-evaluate the fundamental tenets of capitalism. In the words of Nick Hanauer (Hanauer 2018):

> *"The financial crash of 2008 and the cruelly unequal recovery that followed have sparked crises of faith in the three social technologies that define the modern world: democracy, science, and markets. And with human*

> *civilization rushing headlong toward a global climate catastrophe, the timing couldn't be worse."*

The main argument is that many organisations' narrow focus on Return on Shareholders' Funds is ultimately counterproductive, both for society and the organisations themselves.

In their article "Redefining capitalism", Hanauer and his collaborator Eric Beinhocker (2014) state the case clearly:

> *".. elevating the creation of shareholder value to the status of primary objective is based on a faulty assumption that capital is the scarcest resource in an economy, when in reality it's knowledge that's the scarce, critical ingredient in solving problems."*
>
> *"We believe that a reorientation toward seeing businesses as society's problem solvers rather than simply as vehicles for creating shareholder returns would provide a better description of what businesses actually do. It could help executives better balance the interests of the multiple stakeholders they need to manage. It could also help shift incentives back toward long-term investment – after all, few complex human problems can be solved in one quarter."*
>
> *"This is not to say that shareholders or other owners are unimportant. But providing them with a return that is competitive compared with the alternatives is a boundary condition for a successful business; it is not the purpose of a business. After all, having enough food is a boundary condition for life – but the purpose of life is more than just eating."*

The core of their article is an argument about the nature of capitalism and how our view of it needs to be recast in the light of the systemic failures that led to the Global Financial Crisis, in particular the assumption that markets are always 'self-correcting', but the implications for organisations are profound if we believe that how they perceive themselves and their role can make a difference to the way they behave. The authors continue by saying:

> *"If prosperity is created by solving human problems, a key question for society is what kind of economic system will solve the most problems for the most people most quickly. This is the genius of capitalism: it is an unmatched evolutionary system for finding solutions."*
>
> *"Every business is based on an idea about how to solve a problem. The process of converting great ideas into products and services that effectively fulfill fast-changing human needs is what defines most businesses. Thus, the crucial contribution business makes to society is transforming ideas into products and services that solve problems."*

This view not only changes the way that an organisation's performance is measured, it also changes the way the organisation is managed. Adopting Doug Engelbart's vision of Collective IQ provides a powerful means of developing an organisation's problem-solving capabilities, thereby both reflecting and supporting this more enlightened evolution of capitalism. The hope of this book is that it offers a practical framework with which to start implementing this vision.

Appendix I: How to write a Guidance Topic

A Guidance Topic is a much more effective way of communicating policies, procedures, and instructions. The concept optimises use of the networked digital environment and directly supports process management and improvement. Set out below are guidelines on how to write a good Guidance Topic.

As well as being a label for a more effective form of written communication, TOPIC is also an acronym for a structured method to improve writing when working with organisational knowledge. It stands for:

- Targeted
- Organised
- Plain writing
- Illustrated
- Connected

Follow these key words using the guidelines below to create short, sharp, and effective 'atomic units of knowledge'. They will be more relevant, more readily usable, easier to follow, and easier to update and adapt as circumstances change.

Targeted

Before you start writing, check what may already have been written and published about your subject within and outside your organisation. There is no point in reinventing the wheel. Existing internal information may simply need updating and enhancing.

If there are publicly available external documents on the web that cover your subject, use a hyperlink to connect to them from within a shorter document that puts the referenced information into your organisation's context. Don't plagiarize. If hyperlinking isn't possible or practical and you do copy and paste someone else's words, be professional and lawful by using quotes and acknowledging the source.

Also, before you start writing, create a meaningful and specific title. Do this thoughtfully because the title sets your initial 'frame of reference' and is the first step towards achieving clear communication. Use a verb-noun combination to describe a discrete activity or circumstance, for example, "Taking a sales order", or "Preparing the budget", or "Completing an expense claim", or "What to do when a customer complains". Choosing a title is the first test of whether you have a clear idea what you want to communicate and to whom.

Because we often approach writing from the point of the view of an overall subject, rather than with particular people in mind, we often come up with a title that is too large to deal with in one topic.

For example, the title 'Discrimination, harassment, and bullying" covers a clear subject area but does not include a perspective that a reader can immediately relate to. If, however, we write two topics – one titled "What to do if you experience discrimination, bullying, or harassment", aimed at all employees, and one titled "How to respond to a discrimination, harassment, or bullying complaint", aimed at supervisors and managers – then the perspective switches to one that is much more useful. With these issues, an employee's primary concern is what they can do, so the topic is framed specifically to help them. Similarly, supervisors and managers want to know precisely what is expected of them if they receive a complaint.

When you choose titles in this way with the intended purpose and audience implied, there is a much better chance that you will include all the relevant and immediately useful information that the reader needs. And it will be much easier to search for or find when scanning a table of contents.

When you start writing, begin with an introduction that states what the aim of the information is and who will make use of it. For example, an introduction to 'What to do if you are bullied or harassed' might be as follows:

"This topic explains what to do if you feel that you have been bullied or harassed. It provides an explanation of what is meant by bullying and harassment, and then outlines the steps that you can take to get redress and to make sure that it does not continue."

Organised

Structure and write your content. Write down the things that the reader needs to know, in the order that will make most sense in terms of:

- understanding what is being explained, or
- performing steps progressively to complete a task

There are some tricks to make this easier:

- Start by writing a point-form list as an 'outline', and then go back and expand it.
- If a task involves steps by multiple people in different roles, use a 'play-script' layout. This is where you use role names for the 'actors' to identify who does what, and in what order. Here is a simple example:

Marketing Officer

1. Draft text for the product advertising copy.
2. Forward a copy to the Technical Manager with a request to check its technical accuracy.

Technical Manager

3. When you receive a request for a technical check, review the text for technical accuracy.
4. If no changes are required, reply by email that the copy is accurate.
5. If changes are required, make corrections by marking up the supplied copy with suggested changes and/or comments. If major changes are required, contact the Marketing Officer to discuss.
6. Forward the marked-up advertising copy to the Marketing Officer, with further explanations or suggestions that may be helpful.

Marketing Officer

7. When you receive the Technical Manager's reply, make any changes required and then resubmit the copy to the Technical Manager to confirm that it is now technically accurate.
8. Repeat this action until technical accuracy is confirmed.

- Don't be tempted to repeat (or copy and paste) information already covered in other topics. Instead, include hyperlinks so that all topics are parts of an integrated whole.

Plain writing

Edit your draft for plain language. Briefly, this means:

- keep sentences short
- use the active voice
- personalize the information as much as possible
- only use abbreviations, acronyms, or jargon that you are confident readers will understand, otherwise spell things out fully

In particular, it is important to use the active voice. This means writing as if you a speaking directly to the reader. 'Play-script' helps here too because, by identifying the specific people involved (the actors), it makes it easier to address them directly. The result is much clearer and more engaging content.

Good writing is difficult. It takes time, effort, and concentration, so don't expect to be happy with your first draft. More than likely, you will need to rewrite the topic a few times before it reads well. Each time you re-draft, try to clarify and simplify.

Here are some questions to answer:

- Does the title clearly indicate what the topic is about? A good test for this is to ask whether a reader would be able to look at a list of titles and say, "Yes that sounds like the information that I need."
- Does the introduction allow the reader to understand immediately what is covered by the topic and who is involved? If it sounds complicated and convoluted, consider splitting it into several topics, perhaps with an 'overview topic' that links them together.
- Have you used the active voice as much as possible? If you address the reader directly, they are more likely to be interested in what you have to say.
- Are there any terms, abbreviations, or acronyms that the reader may not understand? If so, explain or expand them at their first occurrence, or if there is a central glossary, include a link.

It is extremely valuable to improve your writing skills generally because online written communication is now so integral to working together effectively. There is very good information available on plain writing, both on the internet and in books. Here are some useful references:

- https://www.plainlanguage.gov/guidelines/ (US Government resources published to support the Plain Writing Act of 2010)
- "Writing in Plain English" by Robert D. Eagleson
- "Editing made easy" by Bruce Kaplan

Illustrated

If you can, add examples, diagrams, pictures, or illustrations. But take care to ensure that you explain their relevance in the text and make sure that they fit with the overall communication. Unexplained illustrations that appear seemingly at random may be more confusing than helpful.

Connected

Connect with other topics.

The idea behind creating information in topics is to provide knowledge that, when faced with the need to act or make a decision, is easy to find, easy to understand, and immediately useful.

When you have a collection of topics that cover different aspects of the same general subject, they should form an integrated whole. This means that if a reader needs to understand a given topic in a wider context there should be a link to a topic that covers this.

Conversely, whenever a reader encounters content that assumes knowledge in more specialised detail, or includes contingent activity, there should be links to further topics that give the reader the option of accessing this greater detail.

Appendix II: Guidance Topic vs Policy Document

This appendix takes a deep dive into the differences between taking a conventional approach to an organisational issue and taking a Collective IQ approach. We will use as an example a conventionally constructed document and analyse it to uncover why it is likely to be less effective. Then, we will look at how the issue might be handled using a Guidance Topic approach.

The example we will be using deals with workplace discrimination and harassment – issues that are important to all organisations. The example conventional document is from the Australian Human Rights Commission (AHRC). It is called *Workplace discrimination and harassment policy* and the AHRC makes it available on its public web site as a template. Any organisation is free to copy it and adopt as their own.

Although it is called a template it is presented as a finished document. An organisation can simply copy it, substitute its name for the dummy name 'SureCorp', and then publish it internally. Without any further effort, except perhaps to instruct employees to read it, the organisation can be seen to have fully addressed the issues.

The motives of the AHRC are entirely laudable in that they are trying to help organisations deal with these important issues and comply with the legislation involved. The document itself is not badly written and contains a lot of potentially useful information but it epitomises the shortcomings of this type of approach. Its generic nature and lack of focus on immediate usefulness for intended users (in the context in which they work) severely limits its effectiveness.

This AHRC document is instructive because it is typical of the type written by organisations themselves. It is likely to be constructed in the same way even if it was written from scratch within the organisation. A copy of the document appears on the following pages. After this, we will look closely at its shortcomings and how it can be improved.

SureCorp Ltd

Workplace discrimination and harassment policy

1 Scope

This policy applies to:

- board members
- all staff, including managers and supervisors, full-time, part-time, or casual, temporary, or permanent staff, job candidates, student placements, apprentices, contractors, sub-contractors, and volunteers
- how SureCorp Ltd provides services to clients and how it interacts with other members of the public
- all aspects of employment, recruitment, and selection; conditions and benefits; training and promotion; task allocation; shifts; hours; leave arrangements; workload; equipment and transport
- on-site, off-site, or after-hours work; work-related social functions; conferences – wherever and whenever staff may be as a result of their SureCorp Ltd duties
- staff treatment of other staff, of clients, and of other members of the public encountered in the course of their SureCorp Ltd duties.

2 Aims

SureCorp Ltd is committed to providing a safe, flexible, and respectful environment for staff and clients free from all forms of discrimination, bullying and sexual harassment.

All SureCorp Ltd staff are required to treat others with dignity, courtesy, and respect.

By effectively implementing our Workplace discrimination and harassment policy we will attract and retain talented staff and create a positive environment for staff.

3 Staff rights and responsibilities

All staff are entitled to:

- recruitment and selection decisions based on merit and not affected by irrelevant personal characteristics
- work free from discrimination, bullying, and sexual harassment
- the right to raise issues or to make an enquiry or complaint in a reasonable and respectful manner without being victimised

- reasonable flexibility in working arrangements, especially where needed to accommodate their family responsibilities, disability, religious beliefs, or culture.

All staff must:

- follow the standards of behaviour outlined in this policy
- offer support to people who experience discrimination, bullying or sexual harassment, including providing information about how to make a complaint
- avoid gossip and respect the confidentiality of complaint resolution procedures
- treat everyone with dignity, courtesy, and respect.

3.1 Additional responsibilities of managers and supervisors

Managers and supervisors must also:

- model appropriate standards of behaviour
- take steps to educate and make staff aware of their obligations under this policy and the law
- intervene quickly and appropriately when they become aware of inappropriate behaviour
- act fairly to resolve issues and enforce workplace behavioural standards, making sure relevant parties are heard
- help staff resolve complaints informally
- refer formal complaints about breaches of this policy to the appropriate complaint handling officer for investigation
- ensure staff who raise an issue or make a complaint are not victimised
- ensure that recruitment decisions are based on merit and that no discriminatory requests for information are made
- seriously consider requests for flexible work arrangements.

4 Unacceptable workplace conduct

Discrimination, bullying, and sexual harassment are unacceptable at SureCorp Ltd and are unlawful under the following legislation:

- Sex Discrimination Act 1984 (Cth)
- Racial Discrimination Act 1975 (Cth)
- Disability Discrimination Act 1992 (Cth)
- Age Discrimination Act 2004 (Cth)
- Australian Human Rights Commission Act 1986 (Cth).

Staff (including managers) found to have engaged in such conduct might be counselled, warned, or disciplined. Severe or repeated breaches can lead to formal discipline up to and including dismissal.

4.1 Discrimination

Discrimination is treating, or proposing to treat, someone unfavourably because of a personal characteristic protected by the law, such as sex, age, race, or disability.

Discrimination can occur:

Directly, when a person or group is treated less favourably than another person or group in a similar situation because of a personal characteristic protected by law (see list below).

> For example, a worker is harassed and humiliated because of their race
>
> or
>
> A worker is refused promotion because they are 'too old'

Indirectly, when an unreasonable requirement, condition or practice is imposed that has, or is likely to have, the effect of disadvantaging people with a personal characteristic protected by law (see list below).

> For example, redundancy is decided based on people who have had a worker's compensation claim rather than on merit.

Protected personal characteristics under Federal discrimination law include:

- a disability, disease, or injury, including work-related injury
- parental status or status as a carer, for example, because they are responsible for caring for children or other family members
- race, colour, descent, national origin, or ethnic background
- age, whether young or old, or because of age in general
- sex
- industrial activity, including being a member of an industrial organisation like a trade union or taking part in industrial activity, or deciding not to join a union
- religion
- pregnancy and breastfeeding
- sexual orientation, intersex status or gender identity, including gay, lesbian, bisexual, transsexual, transgender, queer and heterosexual
- marital status, whether married, divorced, unmarried or in a de facto relationship or same sex relationship
- political opinion
- social origin
- medical record
- an association with someone who has, or is assumed to have, one of these characteristics, such as being the parent of a child with a disability.

It is also against the law to treat someone unfavourably because you assume they have a personal characteristic or may have it at some time in the future.

4.2 Bullying

If someone is being bullied because of a personal characteristic protected by equal opportunity law, it is a form of discrimination.

Bullying can take many forms, including jokes, teasing, nicknames, emails, pictures, text messages, social isolation or ignoring people, or unfair work practices.

Under Federal law, this behaviour does not have to be repeated to be discrimination – it may be a one-off event.

Behaviours that may constitute bullying include:

- sarcasm and other forms of demeaning language
- threats, abuse or shouting
- coercion
- isolation
- inappropriate blaming
- ganging up
- constant unconstructive criticism
- deliberately withholding information or equipment that a person needs to do their job or access their entitlements
- unreasonable refusal of requests for leave, training, or other workplace benefits.

Bullying is unacceptable in SureCorp Ltd and may also be against occupational health and safety law.

4.3 Sexual harassment

Sexual harassment is a specific and serious form of harassment. It is unwelcome sexual behaviour, which could be expected to make a person feel offended, humiliated or intimidated. Sexual harassment can be physical, spoken or written. It can include:

- comments about a person's private life or the way they look
- sexually suggestive behaviour, such as leering or staring
- brushing up against someone, touching, fondling or hugging
- sexually suggestive comments or jokes
- displaying offensive screen savers, photos, calendars or objects
- repeated unwanted requests to go out
- requests for sex
- sexually explicit posts on social networking sites
- insults or taunts of a sexual nature
- intrusive questions or statements about a person's private life
- sending sexually explicit emails or text messages

- inappropriate advances on social networking sites
- accessing sexually explicit internet sites
- behaviour that may also be considered to be an offence under criminal law, such as physical assault, indecent exposure, sexual assault, stalking or obscene communications.

Just because someone does not object to inappropriate behaviour in the workplace at the time, it does not mean that they are consenting to the behaviour.

Sexual harassment is covered in the workplace when it happens at work, at work-related events, between people sharing the same workplace, or between colleagues outside of work.

All staff and volunteers have the same rights and responsibilities in relation to sexual harassment.

A single incident is enough to constitute sexual harassment – it doesn't have to be repeated.

All incidents of sexual harassment – no matter how large or small or who is involved – require employers and managers to respond quickly and appropriately.

SureCorp Ltd recognises that comments and behaviour that do not offend one person can offend another. This policy requires all staff and volunteers to respect other people's limits.

4.4 Victimisation

Victimisation is subjecting or threatening to subject someone to a detriment because they have asserted their rights under equal opportunity law, made a complaint, helped someone else make a complaint, or refused to do something because it would be discrimination, sexual harassment or victimisation. Victimisation is against the law.

It is also victimisation to threaten someone (such as a witness) who may be involved in investigating an equal opportunity concern or complaint.

Victimisation is a very serious breach of this policy and is likely (depending on the severity and circumstances) to result in formal discipline against the perpetrator.

SureCorp Ltd has a zero-tolerance approach to victimisation.

4.5 Gossip

It is unacceptable for staff at SureCorp Ltd to talk with other staff members, clients, or suppliers about any complaint of discrimination or harassment.

Breaching the confidentiality of a formal complaint investigation or inappropriately disclosing personal information obtained in a professional role (for example, as a manager) is a serious breach of this policy and may lead to formal discipline.

5 Merit at SureCorp Ltd

All recruitment and job selection decisions at SureCorp Ltd will be based on merit – the skills and abilities of the candidate as measured against the inherent requirements of the position – regardless of personal characteristics.

It is unacceptable and may be against the law to ask job candidates questions, or to in any other way seek information, about their personal characteristics, unless this can be shown to be directly relevant to a genuine requirement of the position.

6 Resolving issues at SureCorp Ltd

SureCorp Ltd strongly encourages any staff member who believes they have been discriminated against, bullied, sexually harassed, or victimised to take appropriate action by contacting their supervisor or manager.

Staff who do not feel safe or confident to take such action may seek assistance from the HR Manager for advice and support or action their behalf.

6.1 Employee assistance program

SureCorp Ltd staff are entitled to a certain amount of free, professional counselling from our employee assistance program. To access the employee assistance program, ring 9999 99999 or email help@assistance.com.

Employee assistance program counselling is confidential, and nothing discussed with a counsellor will be communicated back to SureCorp Ltd. Employee assistance program counselling is available free to SureCorp Ltd staff regardless of whether the issue is related to a workplace problem or some other issue for the staff member.

7 Other relevant SureCorp Ltd policies

Staff, especially managers and supervisors, are encouraged to read this policy in conjunction with other relevant SureCorp Ltd policies, including: list relevant policies.

- Workplace sexual harassment policy
- Flexible work arrangements policy
- Pregnancy and work procedure and policy
- Occupational health and safety policy

- Workplace complaint resolution policy and procedure
- Discipline procedure
- Mission, vision, and values statement
- Enterprise bargaining agreements
- Service agreement

8 More information

If you have a query about this policy or need more information, please contact the HR Manager

9 Review details

This policy was adopted by SureCorp Ltd on 1 April 2018.

This policy was last updated on 1 April 2021

The SureCorp *Workplace discrimination and harassment policy* above is written in a commonly adopted bureaucratic style. It begins by identifying the following groups of people as being subject to the Policy document (under the heading 'Scope'):

- Board Members
- All Staff
  - managers and supervisors
  - full-time, part-time, or casual, temporary, or permanent staff
  - job candidates
  - student placements
  - apprentices
  - contractors
  - sub-contractors
  - volunteers

This is a promising start because it addresses the initial key question of who the information is aimed at. Unfortunately, after identifying them, things start to get a bit confused. There is a later heading for 'Staff rights and responsibilities' and a further section for 'Additional responsibilities of managers and supervisors' but that is the full extent to which the different audiences for the document are addressed directly.

After these headings, the document contains only general subject headings ('Bullying', 'Sexual harassment', 'Victimisation', etc.) that any reader with specific information needs must decipher for themselves. In terms of sorting out what is relevant to whom, the reader is left to pick and choose on their own. This sort of subject-based 'hit and miss' approach can lead to some striking omissions and anomalies. Obvious examples from this document are:

- Board Members: The document doesn't identify any specific action to be taken by Board members. By default, this implies that their obligations are simply the same as Staff.
- Job candidates: This group is bundled in with 'Staff', but they do not have the same relationship with the organisation. Their concern is primarily non-discriminatory recruitment. Other concerns arise only after they join the organisation.
- HR Manager: This position is not mentioned in the document except at the very end under the heading 'More information'. The latter reference implies that the HR Manager plays an important role in handling the issues concerned but the document is silent on precisely what that role is. Logic says the HR Manager's role has the greatest influence on recruitment practices, employment conditions, and training – all of which are critical to handling the issues.
- CEO: The CEO also has a unique and critical role to play but is not separately mentioned. This is the person to whom the Board delegates operational accountability for ensuring that the issues are effectively managed.

This subject-based approach also resulted in another oversight that would not have occurred if a user-centred and topic-based approach had been taken. This is that the document is trying to deal with two issues at the same time that are actually different enough to warrant separate treatments, and therefore separate documents. The two issues are conflated in the document's title but in reality 'discrimination' and 'harassment' are sufficiently different to benefit from separate treatments.

Discrimination mainly concerns denying equal opportunity (of employment, promotion, allocated work, working circumstances, etc.) while harassment (including verbal abuse, unwanted sexual advances, threats, bullying, unfair discrimination, etc.) impacts a person's health and safety.

While harassment may be predicated on sex, race, disability, or age, it may also be indiscriminate. It is primarily simply unacceptable behaviour that is likely to cause mental distress or even physical harm. A more thorough 'target' analysis would have made this obvious.

There was an opportunity to do this towards the beginning of the document under the heading 'Aims'. This is the appropriate place to explain the purpose of the document but disappointingly the 'aims' presented are simply motherhood statements – platitudes[19] that don't cut deeply enough to provide a practical way forward for tackling these issues in the company.

The generic headings 'Aims' and 'Scope' are not helpful either because they separate content that should be combined. The author should be free in the opening paragraphs of the document to clearly state its purpose (Aims) in the context of those involved, and any geographical, jurisdictional, or other limits that apply (Scope). This is what 'targeting' entails and generic standard headings such as 'Aims' and 'Scope' often lead simply to confusion or repetition or both.

Targeted

If a topic-based approach had been taken, targeting the content would have begun by asking the questions 'Who am I talking to?' and 'What do they need to know. We explored this broadly above now let's take a closer look.

Regarding discrimination, the AHRC Policy document talks about 'employment conditions' and 'interactions with clients and the public' but in practice this is overwhelmingly about equal opportunity for job selection and promotion. Discrimination is a societal issue that also manifests in the workplace to unfairly disadvantage some people. An organisation's first priority is to comply with the law and the main tenets of the antidiscrimination laws are:

---

[19] As an example, under "Aims" it says to "treat (people) with dignity, courtesy, and respect" which sounds noble but is not backed up by any commitment to assure it in any way. And while discrimination and harassment may be undignified, discourteous and disrespectful, these sentiments are more appropriate in a general code of conduct where they are at least backed up by the threat of disciplinary action.

- All prohibit discrimination when employing people, promoting them, assigning work, and setting employment conditions.
- All make it an offence to include discriminatory conditions in advertisements.
- The disability and age discrimination acts also make 'victimisation' an offence punishable by up to 6 months imprisonment. Victimisation in this case is threatening or bullying someone who has lodged, or plans to lodge, a complaint with the AHRC.
- The sex discrimination act also prohibits harassment, not just based on sex but also "religious belief, race, colour, or national or ethnic origin". The Disability Discrimination Act also prohibits harassment.

In addition to the federal anti-discrimination laws on sex, race, disability, and age, the Fair Work Act has an anti-discrimination provision, and each State and Territory in Australia has an anti-discrimination or equal opportunity act.

The minimum business need here is that those in charge of the business ensure that employees comply with the law. The broader imperatives are that it is morally wrong to discriminate and there is strong evidence that the lower the degree of discrimination, the better the organisation performs.

Having identified the what, who, and why of the issues, it is not a difficult task to shape this into effective documented business processes.

Organised

We can now combine the 'who' and 'what' by identifying for each 'who', the 'what' that applies. This is summarised in the following table:

Who am I talking to?	**What do they need to know?**
The Board	That they must: ▪ ensure that the organisation complies with anti-discrimination and anti-harassment and legislation. ▪ commit the organisation to operating in a non-discriminatory way and prevent harassment ▪ delegate responsibility to the CEO and approve the resources required ▪ monitor how well the aims are being met by regularly evaluating metrics reported by the CEO
CEO	That they must: ▪ set goals for anti-discrimination and prevention of harassment, including assurance of compliance with legislation ▪ report to the Board on the achievement of goals by collecting and analysing the appropriate metrics ▪ assign responsibilities to the appropriate executives for achieving those goals ▪ deal with discrimination and harassment complaints that are escalated to them
HR Manager	That they must: ▪ design recruitment and appointment mechanisms that avoid discrimination ▪ deal with discrimination complaints that are escalated to them ▪ organise and maintain training in non-discriminatory behaviour by managers and supervisors ▪ measure the effectiveness of anti-discrimination mechanisms and report the results to the CEO.

Who am I talking to?	What do they need to know?
Health & Safety Manager	That they must: ▪ design mechanisms for reporting and dealing with harassment behaviour ▪ organise and maintain training in the meaning, management, and consequences of harassment ▪ measure the effectiveness of harassment prevention mechanisms and report the results to the CEO
Staff (all engagement types)	▪ what it means to be discriminated against ▪ what to do if discriminated against, including if necessary, making a complaint ▪ what to do if a fellow worker is discriminated against ▪ how to respond if a discrimination complaint is made against them ▪ what it means to be harassed and what to do if they are harassed, including making a complaint ▪ how to respond if a harassment complaint is made against them
Managers and supervisors	▪ how to recruit staff, allocate work, recommend promotions, and make other workplace decisions in a non-discriminatory way ▪ what to do if they receive a complaint about discrimination from one of their staff regarding another member of staff ▪ what to do if they receive a complaint about harassment from one of their staff regarding another member of staff
Job candidates	▪ that the organisation has mechanisms in place to ensure that all candidates have an equal opportunity to be successful

Plain English, Illustrated and Connected

Clearly identifying who is involved and what they need to know is a major part of communication in Plain Language. It also points to where best to include content in existing topics and where to create new ones.

Let's start with the perspective of the Board and CEO. They have two issues to deal with, both in terms of policy and practical measures:

- Ensuring that there is no discrimination when recruiting so that everyone enjoys equal opportunities.
- Discouraging harassment and bullying, including that arising from prejudice, to ensure that there is a healthy and positive working environment

For the first issue of discrimination in recruitment, the Board and CEO may agree on, and authorise, a Guidance Topic such as that below. This codifies and communicates the policy on discrimination in recruitment and directs the HR Manager to ensure its practical application and the Company Secretary (or other appropriate executive) to manage legal compliance.

Preventing discrimination in recruitment and employee development

When recruiting for jobs, allocating work, setting working conditions or development opportunities, it is important that we treat everyone equally. Discriminating between people for any reason other than their ability to do the required work harms us because:

- It means that we are not selecting the best candidates and our business performance will suffer.
- It creates a culture of bias that will discourage people from working with us.
- We risk breaking the law. For detailed information on how anti-discrimination law affects businesses, see https://humanrights.gov.au/education/employers.

It is our policy therefore to ensure that discrimination does not occur. To achieve this, do the following:

HR Manager

1. Ensure that the recruitment process includes safeguards against overt discrimination, as well as indirect and unconscious bias.
2. Before a recruitment advertisement is published, ensure that it is checked to remove any potentially discriminatory conditions. Also ensure that it includes advice that the organisation is an equal opportunity employer.

3. When designing the recruitment process, include methods for ensuring that discrimination does not occur and ensure that managers and supervisors are trained in these methods.
4. Devise measures to determine how fair our recruitment practices are and report these six-monthly to the CEO and Board. Include comparisons with measures of diversity in society and industry, in terms of gender, race, age, and disability.
5. When setting employment conditions, ensure that:
   - differing needs are accommodated as far as practicable - these may relate to such things as gender, pregnancy, or childcare, religious or ethnic affiliations, and physical or mental disability.
   - specific groups are not unfairly affected, such as when assigning work or setting rosters
6. If additional expenditure or resources are required above normal operating budgets to carry out these tasks, submit a proposal to the CEO and cite this policy. The CEO will seek Board approval if required.

Company Secretary /Legal Officer /Risk Manager

7. Monitor anti-discrimination and equal opportunity legislation. The Fair Work Act provides for general protection of workers against discrimination and lists all Federal and State Antidiscrimination and Equal Opportunity Acts (see Fair Work Act 2009 / Chapter 3—Rights and responsibilities of employees, employers, organisations etc. / Part 3 1—General protections / Division 5—Other protections / 351 Discrimination).
8. Advise the CEO and Board if changes are made that require changes to our practices.

In response to this topic, the HR Manager can then modify the existing recruitment guidance document (which most organisations will already have) with the following text:

Recruiting and developing staff

… …. …

We aim to maintain a workplace in which all people are treated equally and fairly. All employment related decisions must be based purely on merit. Do not discriminate based on race, colour, sex, sexual orientation, age, physical or mental disability, marital status, family or carer's responsibilities, pregnancy, religion, political opinion, national extraction, or social origin.

When advertising a position, it is illegal to specify any of these characteristics unless they are clearly relevant to being able to do the work. When preparing advertisements for advertising copy, use the templates developed by HR. Before publishing any advertisement or notice, submit it to the HR Manager for review and approval.

Also, when interviewing or communicating with a candidate do not question them on, or discuss, any of these characteristics. You may not take any of these factors into consideration when making your recommendation or decision.

When recruiting for the following positions, use the specified techniques to guard against unconscious bias during the selection process.

- (List positions and associated techniques, for example, blind auditions.)

… … …

To codify policy on harassment and bullying the Board and CEO can use the organisation's existing Code of Conduct. This can also be used to codify the general policy on discrimination (as opposed to the specific key aspect of recruitment). By including it in the Code of Conduct, the policies are addressed directly to all employees and tied-in to existing employee induction, performance monitoring, and disciplinary processes.

Here is additional appropriate text that may be included (note that the title of the Code of Conduct as a Guidance topic is extended to put it in Plain Language).

Code of Conduct - how we expect employees to behave

… … …

Do not discriminate

In any work environment (including all work-related social functions and while travelling on company business) do not discriminate against anyone or harass or make fun of them in any way, based on their race, colour, sex, sexual orientation, age, physical or mental disability, marital status, family or carer's responsibilities, pregnancy, religion, political opinion, national extraction, or social origin. Doing so is contrary to our policy and illegal. Anyone found behaving in this way will be disciplined and may have their employment terminated.

Do not harass or bully others

In any work environment (including all work-related social functions and while travelling on company business) do not engage in harassing or bullying behaviour of any kind. Sexual harassment is a particularly reprehensible and harmful form of behaviour.

Anyone found harassing or bullying others will be disciplined and may have their employment terminated.

Resist discrimination, harassment, and bullying

If you feel that you have been discriminated against, harassed, or bullied in any way, including being unfairly treated or harassed because of your race, colour, sex, sexual orientation, age, physical or mental disability, marital status, family or carer's responsibilities, pregnancy, religion, political opinion, national extraction, or social origin, follow the guidance in the topic *How to respond to discrimination, harassment, or bullying*. This topic includes how to make a formal complaint.

Protect others from discrimination, harassment, and bullying

If you witness discrimination, harassment, or bullying, support your fellow employees by following the guidance in the topic *How to respond to discrimination, harassment, or bullying.*

… …. …

In the text above, the reader is directed to the topic *What to do if you experience discrimination, harassment, or bullying* if they have been subjected to or witnessed any of these behaviours. In an online system this should be connected via a hypertext link so that they can, if they wish, access more detailed information instantly. This Guidance Topic would contain detailed information on how to recognise discrimination, harassment, and bullying and how to go about reporting it.

There would be a further Guidance topic aimed principally at managers and supervisors titled *How to respond to a discrimination, harassment, or bullying complaint*. The latter would include provision for escalating complaint treatment and reporting information on complaints regularly to the CEO and Board.

The HR Manager or Health & Safety Manager would be responsible for authorising and publishing these two Guidance Topics depending on which role is made responsible.

The Collective IQ approach

This example contrasts the single omnibus document in which is dumped a collection of information on a subject and labelled 'policy', with a Collective IQ approach that uses multiple Guidance Topics to meet each organisational actor's knowledge needs.

It replaces a document that constitutes little more than window dressing with a set of documents – Guidance Topics – that support thoroughly considered processes that an organisation can actually have confidence will achieve the stated policy goals for both performance outcomes and legal compliance.

References

Aghina W et al (2017) “The 5 Trademarks of Agile Organisations” McKinsey Quarterly December 2017

AGLS Metadata Standard Part 1 – Reference Description Version 2.0 July 2010, see Term name – function and Term name – documentType

Australian Government Digital Guides - Writing style (https://guides.service.gov.au/content-guide/writing-style/#plain-english) 2022

NSW Premier and Cabinet Plain English Guide (https://www.nsw.gov.au/nsw-government-communications/all-of-government-communications-framework)

Beinhocker E, Hanauer N (2014) “Redefining Capitalism” McKinsey Quarterly September 2014

Bush, Vannevar (1945) "As We May Think", Atlantic Monthly, July 1945

Deming, W. Edwards (1982) “Out of the crisis” Cambridge University Press

Eagleson, Robert D. (1992) “Writing in Plain English” Australian Government Publishing

Eichfeld A. Golding D. Hamilton D. and Robinson K. (2017) “Continuous improvement – make good management every leader’s daily habit” McKinsey & Company

Engelbart, Douglas C (1995) “'Toward augmenting the human intellect and boosting our Collective IQ” Doug Engelbart Institute (http://dougengelbart.org/pubs/books/augment-133150.pdf)

Engelbart, Douglas C (1982) “Toward high-performance knowledge workers”, In OAC '82 Digest, Proceedings of the AFIPS Office Automation Conference. (Apr. 5-7, 1982, San Francisco) 279290. Republished in Computer Supported Cooperative Work: A Book of Readings, Irene Greif, Ed. Morgan Kaufmann, San Mateo, Calif. (1988) 67-78.

Engelbart, Douglas C (1990) "Knowledge-domain interoperability and an open hyperdocument system", In Proceedings of the Conference on Computer-Supported Work. (Oct. 7-10, 1990, Los Angeles) 143-156. Republished in Hypertext/Hypermedia Handbook, E. Berk and J. Devlin, Eds. McGraw-Hill (1991) 397-413.

Engelbart, Douglas C (1992) "Toward high-performance organisations: A strategic role for groupware" In Proceedings of the GroupWare '92 Conference. (Aug. 3-5, 1992, San Jose, Calif.) Morgan Kaufmann, San Mateo, Calif.

Engelbart, Douglas C (1995) "Boosting Our Collective IQ - Selected Readings" Bootstrap Institute/BLT Press

Evans M. Dalkir K. and Bidian C. (2014) "A Holistic View of the Knowledge Life Cycle: The Knowledge Management Cycle (KMC) Model" The Electronic Journal of Knowledge Management Volume 12 Issue 2

Finley, Klint (2018) "50 Years Later, We Still Don't Grasp the Mother of All Demos", Wired Magazine, 12.09.2018' https://www.wired.com/story/50-years-later-we-still-dont-grasp-the-mother-of-all-demos/

Fisher, Adam (2018) "Valley of Genius – The Uncensored History of Silicon Valley" Twelve Hachette Book Group New York First Edition July 2018

Hanauer N (2018) "Capitalism Redefined – Remarks to the OECD " OECD Paris Oct 16 2018

Henisz W. Koller T. and Nuttall R. (2019) "Five ways that ESG creates value" McKinsey Quarterly November 2019

Imai, Masaaki (1986) "Kaizen: The Key to Japan's Competitive Success" McGraw-Hill

Juran, Joseph M. (1951) "Quality Control Handbook" New York, New York: McGraw-Hill

Kaplan, Bruce (2008) "Editing made easy" Penguin Third Edition

Magretta, Joan (2003) "What management is: how it works and why it's everyone's business" Profile Books Ltd

Matthies, Leslie H. (1961) "The playscript procedure: a new tool for administration" Office Publications; Fifth Edition

Nonaka I and Takeuchi H (1995) "The Knowledge-Creating Company: How Japanese Companies Create the Dynamics of Innovation" Oxford University Press

Nonaka I and von Krogh G (2009) "Tacit Knowledge and Knowledge Conversion: Controversy and Advancement in Organisational Knowledge Creation Theory" Organisation Science, Vol. 20, No. 3 (May - Jun 2009), pp. 635-652

Nonaka I and Takeuchi H (1995) "The Knowledge-Creating Company: How Japanese Companies Create the Dynamics of Innovation" Oxford University Press

Ohno, Taiichi (1988) "Toyota Production System: Beyond Large-Scale Production" Productivity Press

O'Mara, Margaret (2018) "In 1968, computers got personal: How the 'mother of all demos' changed the world", The Conversation, 18 September 2018, http://theconversation.com/in-1968-computers-got-personal-how-the-mother-of-all-demos-changed-the-world-101654

Plain Writing Act of 2010 (https://www.govinfo.gov/app/details/PLAW-111publ274)

Porter, Michael E (1985) "'Competitive advantage: Creating and sustaining superior performance', The Free Press

Sandoval A. Lacivita B. Marcos I. and Fookes W. (2019) "Transform the whole business, not just the parts" McKinsey & Company

Schwaber K. Beedle M. (2002). Agile software development with Scrum. Prentice Hall 2002

van Ewyk O. (2016) "Measuring how Well Knowledge is Managed in Organisations" European Conference on Knowledge Management, Academic Conferences International Limited.

van Ewyk O. (2018) "Knowledge Assets and Knowledge Conversion: Addressing issues of practical application" European Conference on Knowledge Management, Academic Conferences International Limited.

Zack, Michael (1999) "Managing Codified Knowledge" Sloan Management Review, Volume 40, Number 4

Made in United States
Orlando, FL
03 November 2026

90134679R00075